Isaac Tod

AF890171

A Treatise on the Integral Calculus and Its Applications

With Numerous Examples

Elibron Classics
www.elibron.com

Elibron Classics series.

© 2005 Adamant Media Corporation.

ISBN 1-4021-0023-X (paperback)
ISBN 1-4021-4598-5 (hardcover)

This Elibron Classics Replica Edition is an unabridged facsimile of the edition published in 1862 by Macmillan and Co., Cambridge and London.

Elibron and Elibron Classics are trademarks of Adamant Media Corporation. All rights reserved.

This book is an accurate reproduction of the original. Any marks, names, colophons, imprints, logos or other symbols or identifiers that appear on or in this book, except for those of Adamant Media Corporation and BookSurge, LLC, are used only for historical reference and accuracy and are not meant to designate origin or imply any sponsorship by or license from any third party.

INTEGRAL CALCULUS.

A TREATISE ON THE

INTEGRAL CALCULUS

AND ITS

APPLICATIONS

WITH NUMEROUS EXAMPLES.

By I. TODHUNTER, M.A.

FELLOW AND PRINCIPAL MATHEMATICAL LECTURER
OF ST JOHN'S COLLEGE, CAMBRIDGE.

SECOND EDITION, REVISED AND ENLARGED.

Cambridge:

MACMILLAN AND CO.

AND 23, HENRIETTA STREET, COVENT GARDEN,

London.

1862.

Cambridge:
PRINTED BY C. J. CLAY, M.A.
AT THE UNIVERSITY PRESS.

PREFACE.

In writing the present treatise on the Integral Calculus, the object has been to produce a work at once elementary and complete—adapted for the use of beginners, and sufficient for the wants of advanced students. In the selection of the propositions, and in the mode of establishing them, I have endeavoured to exhibit fully and clearly the principles of the subject, and to illustrate all their most important results. The process of *summation* has been repeatedly brought forward, with the view of securing the attention of the student to the notions which form the true foundation of the Integral Calculus itself, as well as of its most valuable applications. Considerable space has been devoted to the investigations of the lengths and areas of curves and of the volumes of solids, and an attempt has been made to explain those difficulties which usually perplex beginners—especially with reference to the *limits* of integrations.

The transformation of multiple integrals is one of the most interesting parts of the Integral Calculus, and the experience of teachers shews that the usual modes of treating it are not free from obscurity. I have therefore adopted a method different from those of previous elementary writers, and have endeavoured to render it easily intelligible by full detail, and by the solution of several problems.

The *Calculus of Variations* seems to claim a place in the present treatise with the same propriety as the ordinary theory of maxima and minima values is included in the Differential Calculus. Accordingly the last chapter of the treatise is devoted to this subject; and it is hoped that the theory and illustrations there given will be found, with respect to simplicity and comprehensiveness, adapted to the wants of students.

In order that the student may find in the volume all that he requires, a large collection of examples for exercise has been appended to the several chapters. These examples have been selected from the College and University Examination Papers, and have been carefully verified, so that it is believed that few errors will be found among them.

I. TODHUNTER.

ST JOHN'S COLLEGE,
January, 1862.

CONTENTS.

INTEGRAL CALCULUS.

CHAPTER I.

MEANING OF INTEGRATION. EXAMPLES.

1. IN the Differential Calculus we have a system of rules by means of which we deduce from any given function a second function called the differential coefficient of the former; in the Integral Calculus we have to return from the differential coefficient to the function from which it was deduced. We do not say that this is the *object* of the Integral Calculus, for the fundamental problem of the subject is to effect the summation of a certain infinite series of indefinitely small terms; but for the solution of this problem we must generally know the function of which a given function is the differential coefficient. This we now proceed to shew.

2. Let $\phi(x)$ denote any function of x which remains finite and continuous for all values of x comprised between two fixed values a and b. Let $x_1, x_2, \ldots x_{n-1}$ be a series of values lying between a and b, so that $a, x_1, x_2, \ldots x_{n-1}, b$ are in order of magnitude ascending or descending. We propose then to find the limit of the series

$$(x_1 - a)\,\phi(a) + (x_2 - x_1)\,\phi(x_1) + (x_3 - x_2)\,\phi(x_2) + \ldots\ldots + (b - x_{n-1})\,\phi(x_{n-1}),$$

when $x_1 - a,\ x_2 - x_1, \ldots\ b - x_{n-1}$ are all diminished without limit, and consequently n increased without limit.

Put $x_1 - a = h_1,\ x_2 - x_1 = h_2, \ldots b - x_{n-1} = h_n$; thus the series may be written

$$h_1\,\phi(a) + h_2\,\phi(x_1) \ldots + h_{n-1}\,\phi(x_{n-2}) + h_n\,\phi(x_{n-1}),$$

and may be denoted by $\Sigma h\phi(x)$, for it is the sum of a number of terms of which $h\phi(x)$ may be taken as the type. Since each of the terms of which h is the type may be considered as the difference between two values successively ascribed to the variable x, we may also use the symbol $\phi(x)\,\Delta x$ as the type of the terms to be summed, and $\Sigma\phi(x)\,\Delta x$ for the sum.

We may shew at once that $\Sigma\phi(x)\,\Delta x$ can never exceed a certain finite quantity. For let A denote the numerically greatest value which $\phi(x)$ can have when x lies between a and b; then $\Sigma\phi(x)\,\Delta x$ is numerically less than $(h_1+h_2+\ldots+h_n)\,A$, that is less than $(b-a)\,A$.

We now proceed to determine the limit of $\Sigma\phi(x)\,\Delta x$. Let $\psi(x)$ be such a function of x that $\phi(x)$ is the differential coefficient of it with respect to x. Then we know that the limit of $\dfrac{\psi(x+h)-\psi(x)}{h}$ when h is indefinitely diminished is $\phi(x)$. Hence we may put

$$\psi(x_1)-\psi(a) = h_1\{\phi(a)+\rho_1\},$$
$$\psi(x_2)-\psi(x_1) = h_2\{\phi(x_1)+\rho_2\},$$
$$\cdots\cdots\cdots\cdots\cdots\cdots$$
$$\psi(x_{n-1})-\psi(x_{n-2}) = h_{n-1}\{\phi(x_{n-2})+\rho_{n-1}\},$$
$$\psi(b)-\psi(x_{n-1}) = h_n\{\phi(x_{n-1})+\rho_n\},$$

where $\rho_1, \rho_2, \ldots \rho_n$ ultimately vanish. From these equations we have by addition

$$\psi(b)-\psi(a) = \Sigma\phi(x)\,\Delta x + \Sigma h\rho.$$

Now $\Sigma h\rho$ is less than $(b-a)\,\rho'$ where ρ' denotes the greatest of the quantities $\rho_1, \rho_2, \ldots \rho_n$; hence $\Sigma h\rho$ ultimately vanishes, and we obtain this result, *the limit of $\Sigma\phi(x)\,\Delta x$ when each of the quantities of which Δx is the type diminishes indefinitely is $\psi(b)-\psi(a)$.*

3. The notation used to express the preceding result is

$$\int_a^b \phi(x)\,dx = \psi(b)-\psi(a);$$

the symbol $\int$ is an abbreviation of the word "sum," and dx represents the Δx of $\Sigma\phi(x)\,\Delta x$.

4. Suppose that $h_1, h_2, \ldots h_n$ are all equal; then each of them is equal to $\frac{b-a}{n}$, and x_r is equal to $a + \frac{r}{n}(b-a)$. Hence $\int_a^b \phi(x)\,dx$ is equivalent to the following direction, "divide $b-a$ into n equal parts, each part being h; in $\phi(x)$ substitute for x successively $a, a+h, a+2h, \ldots a+(n-1)h$; add these values together, multiply the sum by h and then diminish h without limit." If these operations are performed we shall have as the result $\psi(b) - \psi(a)$, where $\psi(x)$ is the function of which $\phi(x)$ is the differential coefficient with respect to x.

The student then must carefully observe that for the foundation of the Integral Calculus we have a certain *theorem* and a corresponding *notation*. The *theorem* is the following: let $\psi(x)$ be any function of x, and $\phi(x)$ its differential coefficient with respect to x; let n be a positive integer and $nh = b-a$; then the limit when n is indefinitely increased of

$$h\left\{\phi(a) + \phi(a+h) + \phi(a+2h) + \ldots + \phi(b-h)\right\}$$

is $\psi(b) - \psi(a)$.

The *notation* is that this limit is denoted by $\int_a^b \phi(x)\,dx$,

so that $$\int_a^b \phi(x)\,dx = \psi(b) - \psi(a).$$

As a particular case we may suppose a to be zero; then $nh = b$, and the limit when n is indefinitely increased of

$$h\left\{\phi(0) + \phi(h) + \phi(2h) + \ldots + \phi(b-h)\right\}$$

is denoted by $\int_0^b \phi(x)\,dx$, and is equal to $\psi(b) - \psi(0)$.

5. A single term such as $\phi(x)\,\Delta x$ is frequently called an *element*. It may be observed that the limit of $\Sigma\phi(x)\,\Delta x$ will not be altered in value if we omit a *finite* number of its elements, or add a finite number of similar elements; for

in the limit each element is indefinitely small, and a *finite* number of indefinitely small quantities ultimately vanishes.

6. The above process is called *Integration;* the quantity $\int_a^b \phi(x)\,dx$ is called a *definite integral,* and a and b are called the *limits of the integral.* Since the value of this definite integral is $\psi(b) - \psi(a)$ we must, when a function $\phi(x)$ is to be integrated between assigned limits, first ascertain the function $\psi(x)$ of which $\phi(x)$ is the differential coefficient. To express the connexion between $\phi(x)$ and $\psi(x)$ we have

$$\phi(x) = \frac{d\psi(x)}{dx},$$

and this is also denoted by the equation

$$\int \phi(x)\,dx = \psi(x).$$

In such an equation as the last, where we have no limits assigned, we merely assert that $\psi(x)$ is the function from which $\phi(x)$ can be obtained by differentiation; $\psi(x)$ is here called the *indefinite* integral of $\phi(x)$.

7. The problem of finding the areas of curves was one of those which gave rise to the Integral Calculus, and furnishes an illustration of the preceding articles.

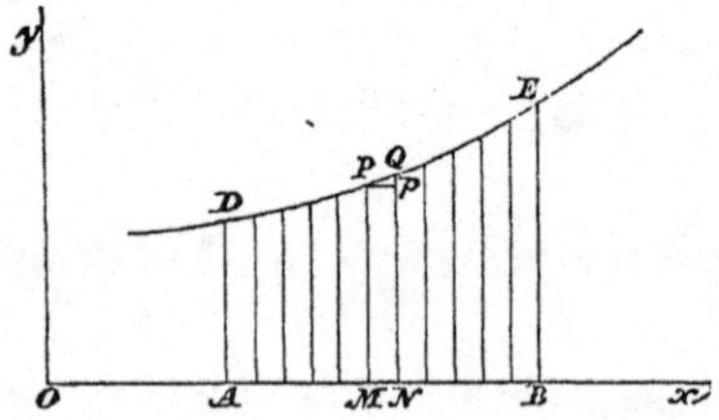

Let DPE be a curve of which the equation is $y = \phi(x)$, and suppose it required to find the area included between this curve, the axis of x, and the ordinates corresponding to the abscissæ a and b. Let $OA = a$, $OB = b$; divide the space

AB into n equal intervals and draw ordinates at the points of division. Suppose $OM = a + (r-1)h$, then the area of the parallelogram $PMNp$ is

$$h\phi\{a + (r-1)h\}.$$

The sum found by assigning to r in this expression all values from 1 to n differs from the required area of the curve by the sum of all the portions similar to the triangle PQp, and as this last sum is obviously less than the greatest of the figures of which $PMNQ$ is one, we can, by sufficiently diminishing h, obtain a result differing as little as we please from the required area. Therefore the area of the curve is the limit of the series

$$h\left\{\phi(a) + \phi(a+h) + \phi(a+2h) + \ldots\ldots + \phi(b-h)\right\},$$

and is equal to $\psi(b) - \psi(a)$.

8. If $\psi(x)$ be the function from which $\phi(x)$ springs by differentiation, we denote this by the equation

$$\int \phi(x)\, dx = \psi(x),$$

and we now proceed to methods of finding $\psi(x)$ when $\phi(x)$ is given. We have shewn, *Dif. Cal.* Art. 102, that if two functions have the same differential coefficient with respect to a variable they can only differ by some constant quantity; hence if $\psi(x)$ be a function having $\phi(x)$ for its differential coefficient with respect to x, then $\psi(x) + C$, where C is any quantity independent of x, is the only form that can have the same differential coefficient. Hence, hereafter, when we assert that any function is the integral of a proposed function, we may if we please add to such integral any constant quantity.

Integration then will for some time appear to be merely the *inverse* of differentiation, and we might have so defined it; we have however preferred to introduce at the beginning the notion of *summation* because it occurs in many of the most important applications of the subject.

We may observe that if $\phi_1(x)$ and $\phi_2(x)$ are any functions of x,

$$\int \{\phi_1(x) + \phi_2(x)\}\, dx = \int \phi_1(x)\, dx + \int \phi_2(x)\, dx;$$

or at least the two expressions which we assert to be equal can only differ by a constant, for if we differentiate both we arrive at the same result, namely, $\phi_1(x) + \phi_2(x)$.

Also, if c be any constant quantity

$$\int c\phi(x)\,dx = c\int \phi(x)\,dx;$$

or at least the two expressions can only differ by a constant.

9. *Immediate integration.*

When a function is recognized to be the differential coefficient of another function we know of course the integral of the first. The following list gives the integrals of the different simple functions;

$$\int x^m\,dx = \frac{x^{m+1}}{m+1}, \qquad \int \frac{dx}{x} = \log x,$$

$$\int a^x\,dx = \frac{a^x}{\log_e a}, \qquad \int e^x\,dx = e^x,$$

$$\int \sin x\,dx = -\cos x, \qquad \int \cos x\,dx = \sin x,$$

$$\int \frac{dx}{\cos^2 x} = \tan x, \qquad \int \frac{dx}{\sin^2 x} = -\cot x,$$

$$\int \frac{dx}{\sqrt{(a^2 - x^2)}} = \sin^{-1}\frac{x}{a} \quad \text{or} \quad = -\cos^{-1}\frac{x}{a},$$

$$\int \frac{dx}{a^2 + x^2} = \frac{1}{a}\tan^{-1}\frac{x}{a} \quad \text{or} \quad = -\frac{1}{a}\cot^{-1}\frac{x}{a}.$$

10. *Integration by substitution.*

The process of integration is sometimes facilitated by substituting for the variable some function of a new variable. Suppose $\phi(x)$ the function to be integrated, and a and b the limits of the integral. It is evident that we may suppose x to be a function of a new variable z, provided that the function chosen is capable of assuming all the values of x required in the integration. Put then $x = f(z)$, and let a' and b' be the values of z, which make $f(z)$ or x equal to a and b respectively; thus $a = f(a')$ and $b = f(b')$. Now suppose that

$\psi(x)$ is the function of which $\phi(x)$ is the differential coefficient, that is suppose $\phi(x) = \frac{d\psi(x)}{dx}$; then

$$\int_a^b \phi(x)\,dx = \psi(b) - \psi(a)$$
$$= \psi\{f(b')\} - \psi\{f(a')\}.$$

But by the principles of the Differential Calculus,

$$\frac{d\psi\{f(z)\}}{dz} = \phi\{f(z)\}f'(z);$$

therefore $$\psi\{f(b')\} - \psi\{f(a')\} = \int_{a'}^{b'} \phi\{f(z)\}f'(z)\,dz$$
$$= \int_{a'}^{b'} \phi(x)\frac{dx}{dz}\,dz;$$

thus $$\int_a^b \phi(x)\,dx = \int_{a'}^{b'} \phi(x)\frac{dx}{dz}\,dz.$$

This result we may write simply thus

$$\int \phi(x)\,dx = \int \phi(x)\frac{dx}{dz}\,dz,$$

provided we remember that when the former integral is taken between certain limits a and b, the latter must be taken between corresponding limits a' and b'.

11. As an example of the preceding article let $\int \frac{dx}{\sqrt{(2ax - x^2)}}$ be required. Assume $x = a - z$, then $\frac{dx}{dz} = -1$, and $2ax - x^2 = a^2 - z^2$. Thus

$$\int \frac{dx}{\sqrt{(2ax - x^2)}} = \int \frac{1}{\sqrt{(2ax - x^2)}}\frac{dx}{dz}\,dz = -\int \frac{dz}{\sqrt{(a^2 - z^2)}}$$
$$= \cos^{-1}\frac{z}{a} = \cos^{-1}\frac{a - x}{a} = \text{vers}^{-1}\frac{x}{a}.$$

Again, let $\int \frac{dx}{x\sqrt{(2ax - a^2)}}$ be required. Assume $x = \frac{a}{1 - z}$, thus

$$\frac{dx}{dz} = \frac{a}{(1 - z)^2}, \text{ and } \int \frac{dx}{x\sqrt{(2ax - a^2)}} = \int \frac{1}{x\sqrt{(2ax - a^2)}}\frac{dx}{dz}\,dz$$

$$=\int \frac{dz}{a\sqrt{\{2(1-z)-(1-z)^2\}}} = \frac{1}{a}\int \frac{dz}{\sqrt{(1-z^2)}}$$

$$=\frac{1}{a}\sin^{-1} z = \frac{1}{a}\sin^{-1}\frac{x-a}{x}.$$

Here we have found the proposed integrals by substituting for x in the manner indicated in the preceding article. This process will often simplify a proposed integral, but no rules can be given to guide the student as to the best assumption to make; this point must be left to observation and practice.

12. *Integration by parts.*

From the equation $\dfrac{d(uv)}{dx} = u\dfrac{dv}{dx} + v\dfrac{du}{dx}$

we deduce by integrating both members,

$$uv = \int u\frac{dv}{dx}\,dx + \int v\frac{du}{dx}\,dx,$$

therefore
$$\int u\frac{dv}{dx}\,dx = uv - \int v\frac{du}{dx}\,dx.$$

The use of this formula is called "integration by parts."

For a particular case suppose $v = x$; then we obtain

$$\int u\,dx = ux - \int x\frac{du}{dx}\,dx.$$

For example, consider $\int x\cos ax\,dx$. Since

$$\cos ax = \frac{1}{a}\frac{d\sin ax}{dx},$$

we may write the proposed expression in the form

$$\int \frac{x}{a}\frac{d\sin ax}{dx}\,dx,$$

and this, by the formula, supposing $u = \dfrac{x}{a}$ and $v = \sin ax$,

$$=\frac{x\sin ax}{a} - \int \frac{\sin ax}{a}\,dx$$

$$= \frac{x \sin ax}{a} - \int \frac{\sin ax}{a} dx$$

$$= \frac{x \sin ax}{a} + \frac{\cos ax}{a^2}.$$

Again, $$\int x^2 \cos ax \, dx = \int \frac{x^2}{a} \frac{d \sin ax}{dx} dx$$

$$= \frac{x^2 \sin ax}{a} - \int \frac{2x}{a} \sin ax \, dx$$

$$= \frac{x^2 \sin ax}{a} + \int \frac{2x}{a^2} \frac{d \cos ax}{dx} dx$$

$$= \frac{x^2 \sin ax}{a} + \frac{2x \cos ax}{a^2} - \int \frac{2 \cos ax}{a^2} dx$$

$$= \frac{x^2 \sin ax}{a} + \frac{2x \cos ax}{a^2} - \frac{2 \sin ax}{a^3}.$$

Again, $$\int e^{cx} \sin ax \, dx = \int \frac{\sin ax}{c} \frac{d\, e^{cx}}{dx} dx$$

$$= \frac{\sin ax}{c} e^{cx} - \int \frac{a e^{cx} \cos ax}{c} dx$$

$$= \frac{\sin ax}{c} e^{cx} - \int \frac{a \cos ax}{c^2} \frac{d\, e^{cx}}{dx} dx$$

$$= \frac{\sin ax}{c} e^{cx} - \frac{a \cos ax}{c^2} e^{cx} - \int \frac{a^2 \sin ax}{c^2} e^{cx} dx.$$

By transposing,

$$\left(1 + \frac{a^2}{c^2}\right) \int e^{cx} \sin ax \, dx = \frac{e^{cx}}{c} \left(\sin ax - \frac{a}{c} \cos ax\right),$$

therefore $$\int e^{cx} \sin ax dx = \frac{e^{cx} (c \sin ax - a \cos ax)}{a^2 + c^2}.$$

Similarly we may shew that

$$\int e^{cx} \cos ax dx = \frac{e^{cx} (c \cos ax + a \sin ax)}{a^2 + c^2}.$$

13. The differential coefficient of any function can always be found by the use of the rules given in the Differential Calculus, but it is not so with the integral of any assigned function. We know, for example, that if m be any number, positive or negative, except -1, then $\int x^m\,dx = \frac{x^{m+1}}{m+1}$, but when $m = -1$ this is not true; in this case we have $\int \frac{dx}{x} = \log x$. If however we had not previously defined the term *logarithm*, and investigated the properties of a *logarithm*, we should have been unable to state what function would give $\frac{1}{x}$ as its differential coefficient. Thus we may find ourselves limited in our powers of integration from our not having given a name to every particular function and investigated its properties.

In order to effect any proposed integration, it will often be necessary to use artifices which can only be suggested by practice.

14. We add a few miscellaneous examples.

Ex. (1). $\int \sqrt{(a^2 - x^2)}\,dx$.

$\int \sqrt{(a^2 - x^2)}\,dx = x\sqrt{(a^2 - x^2)} + \int \frac{x^2\,dx}{\sqrt{(a^2 - x^2)}}$, by Art. 12, supposing $u = \sqrt{(a^2 - x^2)}$ and $v = x$.

And $\int \sqrt{(a^2 - x^2)}\,dx = \int \frac{a^2 - x^2}{\sqrt{(a^2 - x^2)}}\,dx = \int \frac{a^2\,dx}{\sqrt{(a^2 - x^2)}} - \int \frac{x^2\,dx}{\sqrt{(a^2 - x^2)}}$;

therefore, by addition,

$$2\int \sqrt{(a^2 - x^2)}\,dx = x\sqrt{(a^2 - x^2)} + a^2 \int \frac{dx}{\sqrt{(a^2 - x^2)}},$$

therefore $\int \sqrt{(a^2 - x^2)}\,dx = \frac{x\sqrt{(a^2 - x^2)}}{2} + \frac{a^2}{2}\sin^{-1}\frac{x}{a}$. Art. 9.

Ex. (2). $\int \frac{dx}{\sqrt{(x^2 + a^2)}}$.

Assume $\sqrt{(x^2+a^2)} = z - x$, therefore $a^2 = z^2 - 2zx$,

$$\frac{dx}{dz} = \frac{z-x}{z}.$$

Hence $$\int \frac{dx}{\sqrt{(x^2+a^2)}} = \int \frac{1}{\sqrt{(x^2+a^2)}} \frac{dx}{dz} dz = \int \frac{dz}{z} = \log z$$

$$= \log \{x + \sqrt{(x^2+x^2)}\}.$$

Ex. (3). $\displaystyle\int \frac{dx}{\sqrt{(x^2-a^2)}}$.

As in Ex. (2), we may shew that the result is

$$\log \{x + \sqrt{(x^2-a^2)}\}.$$

Ex. (4). $\displaystyle\int \sqrt{(x^2+a^2)}\, dx$.

$$\int \sqrt{(x^2+a^2)}\, dx = x\sqrt{(x^2+a^2)} - \int \frac{x^2 dx}{\sqrt{(x^2+a^2)}} \text{ by Art. 12.}$$

Also $$\int \sqrt{(x^2+a^2)}\, dx = \int \frac{x^2+a^2}{\sqrt{(x^2+a^2)}} dx = \int \frac{x^2 dx}{\sqrt{(x^2+a^2)}} + a^2 \int \frac{dx}{\sqrt{(x^2+a^2)}};$$

therefore, by addition,

$$2\int \sqrt{(x^2+a^2)}\, dx = x\sqrt{(x^2+a^2)} + a^2 \int \frac{dx}{\sqrt{(x^2+a^2)}};$$

therefore $$\int \sqrt{(x^2+a^2)}\, dx = \frac{x\sqrt{(x^2+a^2)}}{2} + \frac{a^2}{2} \log \{x + \sqrt{(x^2+a^2)}\}.$$

Similarly $$\int \sqrt{(x^2-a^2)}\, dx = \frac{x\sqrt{(x^2-a^2)}}{2} - \frac{a^2}{2} \log \{x + \sqrt{(x^2-a^2)}\}.$$

Ex. (5). $\displaystyle\int \frac{dx}{\sqrt{(a+bx+cx^2)}}$.

$$\int \frac{dx}{\sqrt{(a+bx+cx^2)}} = \frac{1}{\sqrt{c}} \int \frac{dx}{\sqrt{\left(\frac{a}{c} + \frac{bx}{c} + x^2\right)}}$$

$$= \frac{1}{\sqrt{c}} \int \frac{dx}{\sqrt{\left\{\left(x + \frac{b}{2c}\right)^2 + \frac{4ac-b^2}{4c^2}\right\}}}.$$

Putting $x+\frac{b}{2c}=z$, our integral becomes, by (2) and (3),

$$\frac{1}{\sqrt{c}}\log\{2cx+b+2\sqrt{c}\sqrt{(a+bx+cx^2)}\},$$

where we omit the constant quantity $\frac{1}{\sqrt{c}}\log 2c$.

In a similar manner, by assuming $z=x+\frac{b}{2c}$ we may make $\int\sqrt{(a+bx+cx^2)}\,dx$ depend upon Ex. (4).

Ex. (6). $\int\frac{dx}{\sqrt{(a+bx-cx^2)}}$.

$$\int\frac{dx}{\sqrt{(a+bx-cx^2)}}=\frac{1}{\sqrt{c}}\int\frac{dx}{\sqrt{\left(\frac{a}{c}+\frac{bx}{c}-x^2\right)}}$$

$$=\frac{1}{\sqrt{c}}\int\frac{dx}{\sqrt{\left\{\frac{4ac+b^2}{4c^2}-\left(x-\frac{b}{2c}\right)^2\right\}}}.$$

Put h^2 for $\frac{4ac+b^2}{4c^2}$ and z for $x-\frac{b}{2c}$, then the integral becomes $\frac{1}{\sqrt{c}}\int\frac{dz}{\sqrt{(h^2-z^2)}}$, which gives $\frac{1}{\sqrt{c}}\sin^{-1}\frac{z}{h}$, or

$$\frac{1}{\sqrt{c}}\sin^{-1}\frac{2cx-b}{\sqrt{(4ac+b^2)}}.$$

In a similar manner, by assuming $z=x-\frac{b}{2c}$ we may make $\int\sqrt{(a+bx-cx^2)}\,dx$ depend upon Ex. (1).

Ex. (7). $\int\frac{dx}{x\sqrt{(x^2-a^2)}}$.

Put $x=\frac{1}{y}$, then $\int\frac{dx}{x\sqrt{(x^2-a^2)}}=\int\frac{1}{x\sqrt{(x^2-a^2)}}\frac{dx}{dy}dy$

$$= -\int \frac{dy}{\sqrt{(1-a^2y^2)}} = -\frac{1}{a}\int \frac{dy}{\sqrt{\left(\frac{1}{a^2}-y^2\right)}} = -\frac{1}{a}\sin^{-1} ay$$

$$= -\frac{1}{a}\sin^{-1}\frac{a}{x}.$$

Since $\sin^{-1}\frac{a}{x} + \cos^{-1}\frac{a}{x} = \frac{\pi}{2}$, a constant, we may also write our last result thus,

$$\int \frac{dx}{x\sqrt{(x^2-a^2)}} = \frac{1}{a}\cos^{-1}\frac{a}{x}.$$

Ex. (8). $\int \frac{dx}{x\sqrt{(a^2 \pm x^2)}}$.

By putting $x = \frac{1}{y}$, as in Ex. (7), we deduce for the required result

$$\frac{1}{a}\log\frac{x}{a+\sqrt{(a^2 \pm x^2)}}.$$

Ex. (9). $\int \frac{dx}{(x-a)^m}$ and $\int \frac{dx}{x-a}$.

$$\int \frac{dx}{(x-a)^m} = -\frac{1}{m-1}\frac{1}{(x-a)^{m-1}},$$

$$\int \frac{dx}{x-a} = \log(x-a).$$

These are obvious if we differentiate the right-hand members.

Ex. 10. $\int \frac{dx}{x^2-a^2}$.

$$\int \frac{dx}{x^2-a^2} = \frac{1}{2a}\int \left(\frac{1}{x-a} - \frac{1}{x+a}\right) dx$$

$$= \frac{1}{2a}\int \frac{dx}{x-a} - \frac{1}{2a}\int \frac{dx}{x+a}$$

$$=\frac{1}{2a}\log(x-a)-\frac{1}{2a}\log(x+a)$$

$$=\frac{1}{2a}\log\frac{x-a}{x+a}.$$

This supposes $\frac{x-a}{x+a}$ positive; if $\frac{x-a}{x+a}$ be negative, we must write

$$\int\frac{dx}{x^2-a^2}=\frac{1}{2a}\log\frac{a-x}{a+x}.$$

Ex. (11). $\int\frac{dx}{a+bx+cx^2}$.

$$\int\frac{dx}{a+bx+cx^2}=\frac{1}{c}\int\frac{dx}{\left(x+\frac{b}{2c}\right)^2+\frac{4ac-b^2}{4c^2}}.$$

If $\frac{4ac-b^2}{4c^2}$ be negative, we obtain the integral by Ex. (10), namely

$$\frac{1}{\sqrt{(b^2-4ac)}}\log\frac{2cx+b-\sqrt{(b^2-4ac)}}{2cx+b+\sqrt{(b^2-4ac)}}.$$

If $\frac{4ac-b^2}{4c^2}$ be positive, then by Art. 9, the integral is

$$\frac{2}{\sqrt{(4ac-b^2)}}\tan^{-1}\frac{2cx+b}{\sqrt{(4ac-b^2)}}.$$

Ex. (12). $\int\frac{Ax+B}{a+bx+cx^2}dx.$

$$\int\frac{Ax+B}{a+bx+cx^2}dx=\int\frac{Ax+\frac{Ab}{2c}+B-\frac{Ab}{2c}}{a+bx+cx^2}dx$$

$$=\frac{A}{2c}\int\frac{2cx+b}{a+bx+cx^2}dx+\left(B-\frac{Ab}{2c}\right)\int\frac{dx}{a+bx+cx^2}.$$

The former integral is $\frac{A}{2c}\log(a+bx+cx^2)$, and the latter has been found in Ex. (11).

Ex. (13). $\int \frac{dx}{\cos x}$.

$$\int \frac{dx}{\cos x} = \int \frac{\cos x\, dx}{\cos^2 x} = \int \frac{dz}{1 - z^2}, \text{ if } z = \sin x,$$

$$= \tfrac{1}{2} \log \frac{1+z}{1-z}, \text{ by Ex. (10)},$$

$$= \tfrac{1}{2} \log \frac{1 + \sin x}{1 - \sin x} = \log \cot \left(\frac{\pi}{4} - \frac{x}{2}\right).$$

Similarly
$$\int \frac{dx}{\sin x} = \log \tan \frac{x}{2}.$$

Ex. (14). $\int \frac{dx}{a + b\cos x}$, and $\int \frac{dx}{a + b \sin x}$.

$$\int \frac{dx}{a + b\cos x} = \int \frac{dx}{a\left(\sin^2\frac{x}{2} + \cos^2\frac{x}{2}\right) + b\left(\cos^2\frac{x}{2} - \sin^2\frac{x}{2}\right)}$$

$$= \int \frac{\sec^2\frac{x}{2}\, dx}{a + b + (a-b)\tan^2\frac{x}{2}}$$

$$= 2\int \frac{dz}{a + b + (a-b)z^2}, \quad \text{if } z = \tan\frac{x}{2}.$$

Hence, if a be greater than b, the integral is

$$\frac{2}{\sqrt{(a^2 - b^2)}} \tan^{-1} \frac{z\sqrt{(a-b)}}{\sqrt{(a+b)}} \text{ or } \frac{2}{\sqrt{(a^2 - b^2)}} \tan^{-1} \frac{\sqrt{(a-b)}\tan\frac{x}{2}}{\sqrt{(a+b)}};$$

and if a be less than b,

$$\frac{1}{\sqrt{(b^2 - a^2)}} \log \frac{z\sqrt{(b-a)} + \sqrt{(b+a)}}{z\sqrt{(b-a)} - \sqrt{(b+a)}},$$

or
$$\frac{1}{\sqrt{(b^2 - a^2)}} \log \frac{\sqrt{(b-a)}\tan\frac{x}{2} + \sqrt{(b+a)}}{\sqrt{(b-a)}\tan\frac{x}{2} - \sqrt{(b+a)}}.$$

To find $\int \frac{dx}{a+b \sin x}$ assume $x = \frac{\pi}{2} + z$; thus the integral becomes $\int \frac{dz}{a+b \cos z}$, which has just been found. Or we may proceed thus,

$$\int \frac{dx}{a+b \sin x} = \int \frac{dx}{a\left(\sin^2 \frac{x}{2} + \cos^2 \frac{x}{2}\right) + 2b \sin \frac{x}{2} \cos \frac{x}{2}}$$

$$= \int \frac{\sec^2 \frac{x}{2}\, dx}{a\left(1 + \tan^2 \frac{x}{2}\right) + 2b \tan \frac{x}{2}}$$

$$= 2\int \frac{dz}{a(1+z^2) + 2bz}, \text{ if } z = \tan \frac{x}{2}.$$

Put $y = z + \frac{b}{a}$, and the integral becomes

$$\frac{2}{a}\int \frac{dy}{y^2 + 1 - \frac{b^2}{a^2}};$$

and this can be found as before.

In any of these examples, since we have found the *indefinite* integral, we can immediately ascertain the definite integral between any assigned limits. For example, since

$$\int \frac{dx}{\sqrt{(x^2+a^2)}} = \log \{x + \sqrt{(x^2+a^2)}\},$$

therefore

$$\int_a^{2a} \frac{dx}{\sqrt{(x^2+a^2)}} = \log [2a + \sqrt{\{(2a)^2 + a^2\}}] - \log \{a + \sqrt{(a^2+a^2)}\}$$

$$= \log \frac{2+\sqrt{5}}{1+\sqrt{2}}.$$

15. The integral $\int x^{m-1} (a+bx^n)^{\frac{p}{q}} dx$ can be found immediately if $\frac{p}{q}$ is a positive integer, for $(a+bx^n)^{\frac{p}{q}}$ can then be expanded by the binomial theorem in a finite series of powers of x, and each term of the product of this series by x^{m-1} will

be immediately integrable. There are also two other cases in which the integral can be found immediately.

For assume $$a + bx^n = t^q;$$

therefore $$x = \left(\frac{t^q - a}{b}\right)^{\frac{1}{n}}, \quad \frac{dx}{dt} = \frac{qt^{q-1}}{nb}\left(\frac{t^q - a}{b}\right)^{\frac{1}{n}-1}.$$

Hence $$\int x^{m-1}(a + bx^n)^{\frac{p}{q}} dx = \int x^{m-1}(a + bx^n)^{\frac{p}{q}} \frac{dx}{dt} dt$$

$$= \frac{q}{nb}\int t^{p+q-1}\left(\frac{t^q - a}{b}\right)^{\frac{m}{n}-1} dt.$$

If $\frac{m}{n}$ *be a positive integer* we can expand $(t^q - a)^{\frac{m}{n}-1}$ in a finite series of powers of t, and each term of the product of this series by t^{p+q-1} will be immediately integrable.

Again, $$\int x^{m-1}(a + bx^n)^{\frac{p}{q}} dx = \int x^{\frac{pn}{q}+m-1}(ax^{-n} + b)^{\frac{p}{q}} dx;$$

and by the former case, if we put $ax^{-n} + b = t^q$, this is immediately integrable if

$$\frac{\frac{pn}{q} + m}{-n}$$

be a positive integer; that is, if $\frac{m}{n} + \frac{p}{q}$ *be a negative integer*.

In the first case, if $\frac{m}{n}$ were a *negative* integer the integral might still be found, as we shall see in the next chapter, and similarly, in the second case, if $\frac{m}{n} + \frac{p}{q}$ were a *positive* integer: but as in these cases some further reductions are necessary, we do not say that the expressions are *immediately* integrable.

Ex. (1). $\int x^2(a + x)^{\frac{1}{2}} dx.$

Here $\frac{m}{n}=3$: assume $a+x=t^2$; the integral becomes

$$2\int(t^2-a)^2t^2dt \text{ or } 2\int(t^6-2at^4+a^2t^2)\,dt,$$

which gives

$$2\left\{\frac{t^7}{7}-\frac{2at^5}{5}+\frac{a^2t^3}{3}\right\};$$

thus $\int x^2(a+x)^{\frac{1}{2}}dx=2\,(a+x)^{\frac{3}{2}}\left\{\frac{(a+x)^2}{7}-\frac{2a}{5}(a+x)+\frac{a^2}{3}\right\}.$

Ex. (2). $\int\frac{dx}{x^2(1+x^2)^{\frac{1}{2}}}.$

Here $m=-1,\quad n=2,\quad \frac{p}{q}=-\frac{1}{2}$;

therefore $\frac{m}{n}+\frac{p}{q}=-1,$

Assume $x^{-2}+1=t^2$;

therefore $x^2=\frac{1}{t^2-1},$

and $\frac{dx}{dt}=-\frac{t}{(t^2-1)^{\frac{3}{2}}}.$

Also $$\int\frac{dx}{x^2(1+x^2)^{\frac{1}{2}}}=\int\frac{\frac{dx}{dt}}{x^3(x^{-2}+1)^{\frac{1}{2}}}dt.$$

Substitute for x and $\frac{dx}{dt}$ their values, and this becomes $-\int dt,$

which $=-t$ or $-\frac{\sqrt{(x^2+1)}}{x}.$

Ex. (3). $\int\frac{dx}{(a^2+x^2)^{\frac{3}{2}}}.$

Here $m=1,\quad n=2,\quad \frac{p}{q}=-\frac{3}{2},$

therefore $$\frac{m}{n}+\frac{p}{q}=-1.$$

Assume $$a^2x^{-2}+1=t^2,$$

therefore $$x^2=\frac{a^2}{t^2-1},$$

$$\frac{dx}{dt}=-\frac{at}{(t^2-1)^{\frac{3}{2}}},$$

$$\int\frac{dx}{(a^2+x^2)^{\frac{3}{2}}}=\int\frac{\frac{dx}{dt}}{x^3(a^2x^{-2}+1)^{\frac{3}{2}}}dt=-\frac{1}{a^2}\int\frac{dt}{t^2}=\frac{1}{a^2t}$$

$$=\frac{x}{a^2\sqrt{(a^2+x^2)}}.$$

EXAMPLES.

1. $\int\frac{dx}{\sqrt{(1-3x-x^2)}}=\sin^{-1}\frac{3+2x}{\sqrt{13}}.$

2. $\int\log x\,dx=x(\log x-1).$

3. $\int x^n\log x\,dx=\frac{x^{n+1}}{n+1}\left\{\log x-\frac{1}{n+1}\right\}.$

4. $\int\theta\sin\theta\,d\theta=-\theta\cos\theta+\sin\theta.$

5. $\int\frac{dx}{e^x+e^{-x}}=\tan^{-1}(e^x).$

6. $\int\sqrt{\left(\frac{m+x}{x}\right)}dx=\sqrt{(mx+x^2)}+m\log\{\sqrt{x}+\sqrt{(m+x)}\}.$

 This may be found by putting $x=z^2$.

7. $\int x\tan^{-1}x\,dx=\frac{1+x^2}{2}\tan^{-1}x-\frac{1}{2}x.$

8. $\int(1-\cos x)^2dx=\frac{3x}{2}-2\sin x+\frac{\sin 2x}{4}.$

9. $\int \frac{x\,dx}{(1-x)^3} = -\frac{1}{1-x} + \frac{1}{2(1-x)^2}.$

10. $\int \frac{x^2\,dx}{a^6-x^6} = \frac{1}{6a^3} \log \frac{a^3+x^3}{a^3-x^3}.$

11. $\int \sqrt{(2ax-x^2)}\,dx = \frac{x-a}{2}\sqrt{(2ax-x^2)} + \frac{a^2}{2}\sin^{-1}\frac{x-a}{a}.$

12. $\int \frac{x\,dx}{\sqrt{(2ax-x^2)}} = -\sqrt{(2ax-x^2)} + a\,\text{vers}^{-1}\frac{x}{a}.$

13. $\int \frac{dx}{\sqrt{(x^2-6x+13)}} = \log\{x-3+\sqrt{(x^2-6x+13)}\}.$

14. $\int \frac{1+\cos x}{x+\sin x}\,dx = \log(x+\sin x).$

15. $\int \frac{x+\sin x}{1+\cos x}\,dx = x\tan\frac{x}{2}.$

16. $\int \frac{dx}{x(\log x)^n} = -\frac{1}{(n-1)(\log x)^{n-1}}.$

17. $\int \frac{\log(\log x)}{x}\,dx = \log x \,.\, \log(\log x) - \log x.$

18. $\int \frac{dx}{x+\sqrt{(x^2-1)}} = \frac{x^2}{2} - \frac{x\sqrt{(x^2-1)}}{2} + \frac{1}{2}\log\{x+\sqrt{(x^2-1)}\}.$

19. $\int \frac{x^3\,dx}{\sqrt{(x-1)}} = 2\sqrt{(x-1)}\left\{\frac{(x-1)^3}{7} + \frac{3}{5}(x-1)^2 + x\right\}.$

20. $$\int e^{ax}\sin mx\cos nx\,dx = \frac{e^{ax}}{2}\,\frac{a\sin(m+n)x-(m+n)\cos(m+n)x}{a^2+(m+n)^2}$$
$$+\frac{e^{ax}}{2}\,\frac{a\sin(m-n)x-(m-n)\cos(m-n)x}{a^2+(m-n)^2}.$$

21. $$\int e^{-x}\cos^3 x\,dx = \frac{1}{4}\int e^{-x}(\cos 3x + 3\cos x)\,dx$$
$$= \frac{e^{-x}}{40}(3\sin 3x - \cos 3x) + \frac{3e^{-x}}{8}(\sin x - \cos x).$$

22. $\int_0^a \sqrt{(a^2 - x^2)}\, dx = \dfrac{\pi a^2}{4}$.

23. $\int_0^{2a} \sqrt{(2ax - x^2)}\, dx = \dfrac{\pi a^2}{2}$.

24. $\int_0^{2a} \operatorname{vers}^{-1} \dfrac{x}{a}\, dx = \pi a$.

Proceed thus; let $\operatorname{vers}^{-1} \dfrac{x}{a} = \theta$, therefore $x = a(1 - \cos\theta)$,

and the integral becomes $\int_0^{\pi} a\theta \sin\theta \, d\theta$.

25. $\int_0^{2a} x \operatorname{vers}^{-1} \dfrac{x}{a}\, dx = \dfrac{5\pi a^2}{4}$.

26. $\int_0^{2a} x^2 \operatorname{vers}^{-1} \dfrac{x}{a}\, dx = \dfrac{11\pi a^3}{6}$.

27. $\int_0^{\frac{1}{2}\pi} \sin^2\theta \cos^3\theta \, d\theta = \frac{2}{15}$.

28. $\int \dfrac{dx}{\sin x + \cos x} = \dfrac{1}{\sqrt{2}} \log \tan\left(\dfrac{x}{2} + \dfrac{\pi}{8}\right)$.

29. $\int \dfrac{dx}{x\sqrt{(a + bx + cx^2)}}$.

Put $x = \dfrac{1}{y}$ and this becomes a known form.

30. $\int \dfrac{\sqrt{(1 - x^2)}}{x^4} \sin^{-1} x dx = -\dfrac{\sin^{-1} x (1 - x^2)^{\frac{3}{2}}}{3x^3} - \dfrac{1}{6x^2} - \dfrac{\log x}{3}$.

This may be obtained by putting $\sin^{-1} x = \theta$.

31. $\int \dfrac{\sin^{-1} x}{(1 - x^2)^{\frac{3}{2}}} dx = \theta \tan\theta + \log\cos\theta$, where $\sin\theta = x$.

32. $\int \dfrac{dx}{(a^2 - x^2)^{\frac{5}{2}}} = \dfrac{1}{a^4}\left(\cot\theta + \dfrac{\cot^3\theta}{3}\right)$, where $x = a\cos\theta$.

33. $\int \dfrac{\sin^2 x \, dx}{a + b\cos^2 x} = \left(\dfrac{a + b}{ab^2}\right)^{\frac{1}{2}} \tan^{-1} \dfrac{\sqrt{a} \tan x}{\sqrt{(a + b)}} - \dfrac{x}{b}$.

34. $\int x^3 \sqrt{(a+bx^2)}\,dx = \left(\frac{a+bx^2}{5b^2} - \frac{a}{3b^2}\right)(a+bx^2)^{\frac{3}{2}}$.

35. $\int \frac{dx}{x^4\sqrt{(1+x^2)}} = \frac{(2x^2-1)\sqrt{(1+x^2)}}{3x^3}$.

36. $\int \tan^{2n}\theta\,d\theta = \frac{x^{2n-1}}{2n-1} - \int \tan^{2n-2}\theta\,d\theta$

$$= \frac{x^{2n-1}}{2n-1} - \frac{x^{2n-3}}{2n-3} + \dots\dots - (-1)^n x + (-1)^n \theta,$$

x being $= \tan\theta$.

37. Shew that $\int_0^\pi \sin mx \sin nx\,dx$ and $\int_0^\pi \cos mx \cos nx\,dx$ are zero if m and n are *unequal integers*, and $= \frac{\pi}{2}$ if m and n are *equal integers*.

38. $\int \left\{\log\left(\frac{x}{a}\right)\right\}^3 dx = x\left\{\log\left(\frac{x}{a}\right)\right\}^3 - 3x\left\{\log\frac{x}{a}\right\}^2 + 6x\log\frac{x}{a} - 6x$.

39. $\int \frac{\cot^{-1}x}{x^2(1+x^2)}\,dx = \frac{\theta^2}{2} - \theta\tan\theta - \log\cos\theta$, where $\cot\theta' = x$.

40. $\int \frac{2a+x}{a+x}\sqrt{\left(\frac{a-x}{a+x}\right)}\,dx = \sqrt{(a^2-x^2)} - \frac{2a\sqrt{(a-x)}}{\sqrt{(a+x)}}$.

41. $\int \frac{\text{vers}^{-1}\frac{x}{a}}{\sqrt{(2ax-x^2)}}\,dx = \frac{1}{2}\left(\text{vers}^{-1}\frac{x}{a}\right)^2$.

42. $\int_0^{\frac{1}{2}\pi} \frac{dx}{1+c\cos x} = \frac{1}{\sqrt{(1-c^2)}}\cos^{-1}c$, if c is less than 1.

43. $\int_{-\frac{1}{2}\pi}^{\frac{1}{2}\pi} e^{-\theta}\cos^3\theta\,d\theta = \frac{3}{10}\left(e^{\frac{1}{2}\pi} + e^{-\frac{1}{2}\pi}\right)$.

44. $\int \frac{(x^2-1)\,dx}{x\sqrt{(1+3x^2+x^4)}}$. Assume $z = x + \frac{1}{x}$.

45. $\int \frac{(a+bx^n)^{\frac{3}{4}}\,dx}{x}$. Assume $a + bx^n = z^4$.

CHAPTER II.

RATIONAL FRACTIONS.

16. We proceed to the integration of such expressions as

$$\frac{A' + B'x + C'x^2 \ldots + M'x^m}{A + Bx + Cx^2 \ldots + Nx^n},$$

where A, B,...A', B',... are constants, so that both numerator and denominator are finite rational functions of x. If m be equal to n, or greater than n, we may by division reduce the preceding to the form of an integral function of x, and a fraction in which the numerator is of lower dimensions in x than the denominator. As the integral function of x can be integrated immediately, we may confine ourselves to the case of a fraction having its numerator at least one dimension lower than its denominator. In order to effect the integration we resolve the fraction into a series of more simple fractions called *partial fractions*, the possibility of which we proceed to demonstrate.

Let $\frac{U}{V}$ be a rational fraction in its lowest terms which is to be resolved into a series of partial fractions; suppose V a function of x of the nth degree, and U a function of x of the $(n-1)^{\text{th}}$ degree at most; we may without loss of generality suppose the coefficient of x^n in V to be unity. Suppose

$$V = (x-a)(x-b)^r (x^2 - 2\alpha x + \alpha^2 + \beta^2)(x^2 - 2\gamma x + \gamma^2 + \delta^2)^s,$$

so that the equation $V = 0$ has

(1) one real root $= a$,

(2) r equal real roots, each $= b$,

(3) a pair of imaginary roots $\alpha \pm \beta\sqrt{(-1)}$,

(4) s pairs of imaginary roots, each being $\gamma \pm \delta\sqrt{(-1)}$.

By the theory of equations V must be the product of factors of the form we have supposed, the factors being more or fewer in number. Since V is of the nth degree we have

$$1 + r + 2 + 2s = n.$$

Assume

$$\frac{U}{V} = \frac{A}{x-a} + \frac{B_1}{(x-b)^r} + \frac{B_2}{(x-b)^{r-1}} + \frac{B_3}{(x-b)^{r-2}} \dots\dots + \frac{B_r}{x-b}$$

$$+ \frac{Cx+D}{x^2 - 2\alpha x + \alpha^2 + \beta^2}$$

$$+ \frac{E_1 x + F_1}{(x^2 - 2\gamma x + \gamma^2 + \delta^2)^s} + \frac{E_2 x + F_2}{(x^2 - 2\gamma x + \gamma^2 + \delta^2)^{s-1}} \dots + \frac{E_s x + F_s}{x^2 - 2\gamma x + \gamma^2 + \delta^2},$$

where A, B_1, B_2, ... C, D, E_1, ... are constants which, in order to justify our assumption, we must shew can be so determined as to make the second member of the above equation *identically* equal to the first. If we bring all the partial fractions to a common denominator and add them together, we have V for that common denominator, and for the numerator a function of x of the $(n-1)^{\text{th}}$ degree. If we equate the coefficients of the different powers of x in this numerator with the corresponding coefficients in U, we shall have *n equations of the first degree to determine the n quantities* A, B_1, B_2, ... and with these values of A, B_1, B_2, ... the second member of the above equation becomes *identically equal* to the first, and thus $\frac{U}{V}$ is decomposed into a series of partial fractions.

If V involves other single factors like $x - a$, each such factor will give rise to a fraction like $\frac{A}{x-a}$, and any repeated factor like $(x-b)^r$ will give rise to a series of partial fractions of the form $\frac{B_1}{(x-b)^r}$, $\frac{B_2}{(x-b)^{r-1}}$, &c. In like manner other factors of the form $x^2 - 2\alpha x + \alpha^2 + \beta^2$ or $(x^2 - 2\gamma x + \gamma^2 + \delta^2)^s$ will give rise to a fraction or a series of fractions respectively of the forms indicated above.

17. The demonstration given in Art. 16 is not very satisfactory, since we have not proved that the n equations of the

first degree which we use to determine A, B_1, B_2,... are *independent* and *consistent*.

A method of greater rigour has been given in a treatise on the Integral Calculus by Mr Homersham Cox, which we will here briefly indicate. Suppose $F(x)$ to contain the factor $x-a$ repeated n times; we have, if

$$F(x) = (x-a)^n \psi(x),$$

$$\frac{\phi(x)}{F(x)} = \frac{\phi(x)}{(x-a)^n \psi(x)} = \frac{\phi(x) - \dfrac{\phi(a)}{\psi(a)}\psi(x)}{(x-a)^n \psi(x)} + \frac{\dfrac{\phi(a)}{\psi(a)}}{(x-a)^n}.$$

Now $\phi(x) - \dfrac{\phi(a)}{\psi(a)}\psi(x)$ vanishes when $x=a$, and is therefore divisible by $x-a$; suppose the quotient denoted by $\chi(x)$, then

$$\frac{\phi(x)}{F(x)} = \frac{\chi(x)}{(x-a)^{n-1}\psi(x)} + \frac{\phi(a)}{\psi(a)} \frac{1}{(x-a)^n}.$$

The process may now be repeated on $\dfrac{\chi(x)}{(x-a)^{n-1}\psi(x)}$, and thus by successive operations the decomposition of $\dfrac{\phi(x)}{F(x)}$ completely effected. In this proof a may be either a real root or an imaginary root of the equation $F(x)=0$; if $a=\alpha+\beta\sqrt{(-1)}$, then $\alpha-\beta\sqrt{(-1)}$ will also be a root of $F(x)=0$; let b denote this root, then if we add the two partial fractions

$$\frac{\phi(a)}{\psi(a)} \frac{1}{(x-a)^n} \text{ and } \frac{\phi(b)}{\psi(b)} \frac{1}{(x-b)^n},$$

we shall obtain a result free from $\sqrt{(-1)}$.

18. With respect to the integration of these partial fractions we refer to Examples (9) and (12) of Art. 14 for all the forms except $\dfrac{Lx+M}{(x^2-2\gamma x+\gamma^2+\delta^2)^m}$, and this will be given hereafter.

Having proved that a rational fraction can be decomposed in the manner assumed in Art. 16, we may make use of different algebraical artifices in order to diminish the labour of determining A, B_1, B_2, &c. The most useful consideration is, that since the numerator of the proposed fraction is *identically* equal to the numerator formed by adding together the partial fractions, if we assign *any* value to the variable x the equality still subsists.

19. *To determine the partial fraction corresponding to a single factor of the first degree.*

Suppose $\dfrac{\phi(x)}{F(x)}$ represents a fraction to be decomposed, and let $F(x)$ contain the factor $x-a$ once; assume

$$\frac{\phi(x)}{F(x)} = \frac{A}{x-a} + \frac{\chi(x)}{\psi(x)} \quad \text{.....................(1)},$$

where A is a constant, and $\dfrac{\chi(x)}{\psi(x)}$ represents the sum of all the partial fractions exclusive of $\dfrac{A}{x-a}$, and $F(x)=(x-a)\,\psi(x)$.

From (1)

$$\phi(x) = A\psi(x) + (x-a)\,\chi(x) \quad \text{..............(2)}.$$

In (2), which holds for any value of x, make $x=a$, then

$$\phi(a) = A\psi(a),$$

therefore

$$A = \frac{\phi(a)}{\psi(a)}.$$

Since $F'(x) = \psi(x) + (x-a)\,\psi'(x)$, we have

$$F'(a) = \psi(a),$$

therefore

$$A = \frac{\phi(a)}{F'(a)}.$$

20. *To determine the partial fractions corresponding to a factor of the first degree which is repeated.*

Suppose $F(x)$ contains a factor $x-a$ repeated n times, and let $F(x) = (x-a)^n\,\psi(x)$. Assume

$$\frac{\phi(x)}{F(x)} = \frac{A_1}{(x-a)^n} + \frac{A_2}{(x-a)^{n-1}} + \frac{A_3}{(x-a)^{n-2}} \cdots\cdots + \frac{A_n}{x-a} + \frac{\chi(x)}{\psi(x)},$$

where $\frac{\chi(x)}{\psi(x)}$ denotes the sum of the partial fractions arising from the other factors of $F(x)$. Multiply both sides of the equation by $(x-a)^n$ and put $f(x)$ for $\frac{\phi(x)}{F(x)}(x-a)^n$; thus

$$f(x) = A_1 + A_2(x-a) + A_3(x-a)^2 \ldots + A_n(x-a)^{n-1} + \frac{\chi(x)}{\psi(x)}(x-a)^n.$$

Differentiate successively both members of this identity and put $x=a$ *after* differentiation; then

$$f(a) = A_1,$$

$$f'(a) = A_2,$$

$$f''(a) = 1.2A_3,$$

$$f'''(a) = \lfloor 3 A_4,$$

$$\cdots\cdots\cdots\cdots\cdots$$

$$f^{n-1}(a) = \lfloor n-1 A_n.$$

Thus A_1, A_2,... A_n are determined.

21. *To determine the partial fractions corresponding to a pair of imaginary roots which do not recur.*

Let $\frac{\phi(x)}{F(x)}$ denote the fraction to be decomposed; and $\alpha \pm \beta\sqrt{(-1)}$ a pair of imaginary roots; then if we denote these roots by a and b and proceed as in Art. 19, we have for the partial fractions

$$\frac{\phi(a)}{F'(a)}\frac{1}{x-a} \text{ and } \frac{\phi(b)}{F'(b)}\frac{1}{x-b}.$$

Suppose $\frac{\phi(a)}{F'(a)} = A - B\sqrt{(-1)}$; then since $\frac{\phi(b)}{F'(b)}$ may be obtained from $\frac{\phi(a)}{F'(a)}$ by changing the sign of $\sqrt{(-1)}$, we must have $\frac{\phi(b)}{F'(b)} = A + B\sqrt{(-1)}$. Hence the fractions are

$$\frac{A - B\sqrt{(-1)}}{x - \alpha - \beta\sqrt{(-1)}} \text{ and } \frac{A + B\sqrt{(-1)}}{x - \alpha + \beta\sqrt{(-1)}};$$

and their sum is

$$\frac{2A(x-\alpha) + 2B\beta}{(x-\alpha)^2 + \beta^2}.$$

22. Or we may proceed thus. Suppose $x^2 - px + q$ to denote the quadratic factor which gives rise to the pair of imaginary roots $\alpha \pm \beta\sqrt{(-1)}$; then assume

$$\frac{\phi(x)}{F(x)} = \frac{Lx + M}{x^2 - px + q} + \frac{\chi(x)}{\psi(x)},$$

so that $F(x) = (x^2 - px + q)\,\psi(x)$. Multiply by $F(x)$; thus

$$\phi(x) = (Lx + M)\,\psi(x) + (x^2 - px + q)\,\chi(x) \dots\dots\dots (1).$$

Now ascribe to x either of the values which makes $x^2 - px + q$ vanish; then (1) reduces to

$$\phi(x) = (Lx + M)\,\psi(x) \dots\dots\dots\dots\dots\dots (2).$$

Now by the repeated substitution of $px - q$ for x^2 in both members of (2), we shall at last have x occurring in the first power only, so that the equation takes the form

$$Px + Q = P'x + Q'.$$

Now put for x its value $\alpha + \beta\sqrt{(-1)}$ and equate the coefficients of the impossible parts; thus

$$P = P' \text{ and therefore also } Q = Q'.$$

Here P and Q are known quantities, and P' and Q' involve the unknown quantities L and M to the first power only, so that we have two equations of the first degree for finding L and M.

23. *To determine the partial fractions corresponding to a pair of imaginary roots which is repeated.*

We may proceed as in Art. 20. Or we may adopt the following method. Suppose $x^2 - px + q$ to be the quadratic factor which occurs r times; assume

$$\frac{\phi(x)}{F(x)} = \frac{L_r x + M_r}{(x^2 - px + q)^r} + \frac{L_{r-1} x + M_{r-1}}{(x^2 - px + q)^{r-1}} \ldots\ldots + \frac{L_1 x + M_1}{x^2 - px + q} + \frac{\chi(x)}{\psi(x)},$$

so that $F(x) = (x^2 - px + q)^r \psi(x)$. Multiply by $F(x)$; thus

$$\phi(x) = (L_r x + M_r)\,\psi(x) + (L_{r-1} x + M_{r-1})(x^2 - px + q)\,\psi(x)$$
$$+ \ldots\ldots + (x^2 - px + q)^r \chi(x) \ldots\ldots\ldots\ldots (1).$$

Now ascribe to x either of the values which makes $x^2 - px + q$ vanish; thus the equation reduces to

$$\phi(x) = (L_r x + M_r)\,\psi(x).$$

Proceed as in Art. 22, and thus find L_r and M_r. Then from (1) by transposition we have

$$\phi(x) - (L_r x + M_r)\,\psi(x) = (L_{r-1} x + M_{r-1})(x^2 - px + q)\,\psi(x) + \ldots$$

The right-hand member has $x^2 - px + q$ for a factor of every term; hence as the two members are *identical* we can divide by this factor. Let $\phi_1(x)$ indicate the quotient obtained on the left; then

$$\phi_1(x) = (L_{r-1} x + M_{r-1})\,\psi(x) + (L_{r-2} x + M_{r-2})(x^2 - px + q)\,\psi(x)$$
$$+ \ldots\ldots + (x^2 - px + q)^{r-1} \chi(x) \ldots\ldots\ldots\ldots (2).$$

From (2) we find L_{r-1} and M_{r-1} as before; then by transposition and division

$$\phi_2(x) = (L_{r-2} x + M_{r-2})\,\psi(x) + (L_{r-3} x + M_{r-3})(x^2 - px + q)\,\psi(x) + \ldots$$

and so on until all the quantities are determined.

24. Take for example $\dfrac{x^2 - 3x - 2}{(x^2 + x + 1)^2 (x + 1)^2}$. Assume it equal to

$$\frac{L_2 x + M_2}{(x^2 + x + 1)^2} + \frac{L_1 x + M_1}{x^2 + x + 1} + \frac{\chi(x)}{(x + 1)^2};$$

then $x^2 - 3x - 2 = (L_2 x + M_2)(x + 1)^2$

$$+ (L_1 x + M_1)(x^2 + x + 1)(x + 1)^2 + (x^2 + x + 1)^2 \chi(x) \ldots\ldots (1).$$

Suppose $x^2+x+1=0$; thus the equation reduces to

$$x^2-3x-2=(L_2x+M_2)(x+1)^2$$
$$=(L_2x+M_2)(x^2+2x+1).$$

Put $-x-1$ for x^2; thus

$$-4x-3=(L_2x+M_2)x=L_2x^2+M_2x$$
$$=-L_2(x+1)+M_2x;$$

therefore $\qquad -4=-L_2+M_2$, and $-3=-L_2$;

thus $\qquad L_2=3$, and $M_2=-1$.

From (1) by transposition

$$x^2-3x-2-(3x-1)(x+1)^2$$
$$=(L_1x+M_1)(x^2+x+1)(x+1)^2+(x^2+x+1)^2\chi(x).$$

The left-hand member is $-3x^3-4x^2-4x-1$; divide by x^2+x+1; thus

$$-(3x+1)=(L_1x+M_1)(x+1)^2+(x^2+x+1)\chi(x)\text{......}(2).$$

Again, suppose $x^2+x+1=0$; thus

$$-3x-1=(L_1x+M_1)(x^2+2x+1)=(L_1x+M_1)x$$
$$=-L_1(x+1)+M_1x;$$

therefore $\qquad -3=-L_1+M_1$, and $-1=-L_1$;

thus $\qquad L_1=1$ and $M_1=-2$.

Thus the partial fractions corresponding to the quadratic factor are found. The partial fractions corresponding to the factor $(x+1)^2$ may then be found by Art. 20. Or we may from (2) by transposition and division by x^2+x+1 obtain

$$-(x-1)=\chi(x).$$

Thus

$$\frac{\chi(x)}{(x+1)^2}=-\frac{x-1}{(x+1)^2}=-\frac{x+1}{(x+1)^2}+\frac{2}{(x+1)^2}=-\frac{1}{x+1}+\frac{2}{(x+1)^2};$$

therefore

$$\frac{x^2-3x-2}{(x^2+x+1)^2(x+1)^2}=\frac{3x-1}{(x^2+x+1)^2}+\frac{x-2}{x^2+x+1}+\frac{2}{(x+1)^2}-\frac{1}{x+1}.$$

25. Examples. Required the integral of $\frac{5x^3+1}{x^2-3x+2}$.

By division we have

$$\frac{5x^3+1}{x^2-3x+2}=5x+15+\frac{35x-29}{x^2-3x+2}.$$

Assume $$\frac{35x-29}{x^2-3x+2}=\frac{A}{x-1}+\frac{B}{x-2};$$

therefore $$35x-29=A(x-2)+B(x-1).$$

Make x successively equal to 1 and 2; then

$$35-29=-A, \text{ or } A=-6,$$

$$70-29=\quad B, \text{ or } B=\quad 41;$$

therefore $$\frac{5x^3+1}{x^2-3x+2}=5x+15-\frac{6}{x-1}+\frac{41}{x-2};$$

therefore $$\int\frac{5x^3+1}{x^2-3x+2}\,dx=\frac{5x^2}{2}+15x-6\log(x-1)+41\log(x-2).$$

Required the integral of $\frac{9x^2+9x-128}{x^3-5x^2+3x+9}$.

Since $x^3-5x^2+3x+9=(x-3)^2(x+1)$, we assume

$$\frac{9x^2+9x-128}{x^3-5x^2+3x+9}=\frac{A}{x+1}+\frac{B_1}{(x-3)^2}+\frac{B_2}{x-3};$$

therefore $$9x^2+9x-128=A(x-3)^2+B_1(x+1)+B_2(x+1)(x-3).$$

Make $x=3$ and -1 successively, and we find

$$B_1=-5, \quad A=-8.$$

Also by equating the coefficients of x^2, we have

$$9=A+B_2,$$

therefore $$B_2=17;$$

therefore

$$\int\frac{9x^2+9x-128}{x^3-5x^2+3x+9}\,dx=-8\log(x+1)+\frac{5}{x-3}+17\log(x-3).$$

Required the integral of $\frac{x^2+1}{(x-1)^4(x^3+1)}$.

Assume $\frac{x^2+1}{(x-1)^4(x^3+1)}$

$$=\frac{A_1}{(x-1)^4}+\frac{A_2}{(x-1)^3}+\frac{A_3}{(x-1)^2}+\frac{A_4}{x-1}+\frac{B}{x+1}+\frac{Cx+D}{x^2-x+1};$$

therefore $x^2+1=\{A_1+A_2(x-1)+A_3(x-1)^2+A_4(x-1)^3\}(x^3+1)$

$$+\{B(x^2-x+1)+(Cx+D)(x+1)\}(x-1)^4 \ldots (1).$$

Put $x=1$, then $\quad 2=2A_1 \ldots\ldots\ldots\ldots\ldots\ldots (2)$;

therefore $\quad A_1=1$.

From (1) and (2) we have by subtraction,

$$x^2-1=A_1(x^3-1)+\{A_2+A_3(x-1)+A_4(x-1)^2\}(x-1)(x^3+1)$$
$$+\{B(x^2-x+1)+(Cx+D)(x+1)\}(x-1)^4.$$

Divide by $x-1$, then

$$x+1=A_1(x^2+x+1)+\{A_2+A_3(x-1)+A_4(x-1)^2\}(x^3+1)$$
$$+\{B(x^2-x+1)+(Cx+D)(x+1)\}(x-1)^3 \ldots (3).$$

Put $x=1$, then $\quad 2=3A_1+2A_2 \ldots\ldots\ldots\ldots\ldots\ldots (4)$;

therefore $\quad A_2=-\frac{1}{2}$.

From (3) and (4), by subtraction,

$$x-1=A_1(x^2+x-2)+A_2(x^3-1)+\{A_3+A_4(x-1)\}(x-1)(x^3+1)$$
$$+\{B(x^2-x+1)+(Cx+D)(x+1)\}(x-1)^3.$$

Divide by $x-1$, then

$$1=A_1(x+2)+A_2(x^2+x+1)+\{A_3+A_4(x-1)\}(x^3+1)$$
$$+\{B(x^2-x+1)+(Cx+D)(x+1)\}(x-1)^2 \ldots (5).$$

Put $x=1$, then $\quad 1=3A_1+3A_2+2A_3$..............(6);

therefore $\quad A_3=-\frac{1}{4}$.

From (5) and (6), by subtraction,

$$0=A_1(x-1)+A_2(x^2+x-2)+A_3(x^3-1)+A_4(x-1)(x^3+1)$$
$$+\{B(x^2-x+1)+(Cx+D)(x+1)\}(x-1)^2.$$

Divide by $x-1$, then

$$0=A_1+A_2(x+2)+A_3(x^2+x+1)+A_4(x^3+1)$$
$$+\{B(x^2-x+1)+(Cx+D)(x+1)\}(x-1) \quad \text{......(7)}.$$

Put $x=1$, then $0=A_1+3A_2+3A_3+2A_4$............(8);

therefore $\quad A_4=\frac{5}{8}$.

From (7) and (8), by subtraction,

$$0=A_2(x-1)+A_3(x^2+x-2)+A_4(x^3-1)$$
$$+\{B(x^2-x+1)+(Cx+D)(x+1)\}(x-1).$$

Divide by $x-1$, then

$$0=A_2+A_3(x+2)+A_4(x^2+x+1)$$
$$+B(x^2-x+1)+(Cx+D)(x+1) \quad \text{.........(9)}.$$

Put $x=-1$, then

$$0=A_2+A_3+A_4+3B \quad \text{....................(10)};$$

therefore $\quad B=\frac{1}{24}$.

From (9) and (10), by subtraction,

$$0=A_3(x+1)+A_4(x^2+x)+B(x^2-x-2)+(Cx+D)(x+1).$$

Divide by $x+1$, then

$$0=A_3+A_4x+B(x-2)+Cx+D \quad \text{...........(11)}.$$

Put $x=0$, then

$$A_3-2B+D=0 \quad \text{....................(12)};$$

therefore $\quad D=\frac{1}{3}$.

From (11) and (12), by subtraction,

$$A_4 + B + C = 0;$$

therefore $$C = -\tfrac{2}{3};$$

therefore $$\frac{x^2+1}{(x-1)^4(x^3+1)} = \frac{1}{(x-1)^4} - \frac{1}{2(x-1)^3} - \frac{1}{4(x-1)^2}$$

$$+ \frac{5}{8(x-1)} + \frac{1}{24(x+1)} - \frac{2x-1}{3(x^2-x+1)};$$

therefore $$\int \frac{(x^2+1)\,dx}{(x-1)^4(x^3+1)} = -\frac{1}{3(x-1)^3} + \frac{1}{4(x-1)^2} + \frac{1}{4(x-1)}$$

$$+ \frac{5}{8}\log(x-1) + \frac{1}{24}\log(x+1) - \frac{1}{3}\log(x^2-x+1).$$

26. We will give as additional examples the integration of $\dfrac{x^{m-1}}{x^n \pm 1}$, supposing m and n positive integers, and $m-1$ less than n.

Required the integral of $\dfrac{x^{m-1}}{x^n - 1}$, n being supposed even.

By the theory of equations the real roots of $x^n - 1 = 0$ are 1 and -1, and the imaginary roots are found from the expression $\cos r\theta \pm \sqrt{(-1)} \sin r\theta$, where $\theta = \dfrac{\pi}{n}$, and r takes in succession the values 2, 4, ... up to $n-2$. Now by Art. 19 if $\dfrac{\phi(x)}{F(x)}$ be the fraction to be decomposed, the partial fraction corresponding to the root a is $\dfrac{\phi(a)}{F'(a)} \dfrac{1}{x-a}$. In the present case

$$\frac{\phi(a)}{F'(a)} = \frac{a^{m-1}}{na^{n-1}} = \frac{a^m}{na^n} = \frac{a^m}{n}, \text{ since } a^n = 1.$$

Hence corresponding to the root 1 we have the partial fraction $\dfrac{1}{n(x-1)}$, and corresponding to the root -1 we have the partial fraction $-\dfrac{(-1)^m}{n(x+1)}$. And corresponding to the pair of roots

$$\cos r\theta \pm \sqrt{(-1)} \sin r\theta$$

we have

$$\frac{\{\cos r\theta + \sqrt{(-1)}\sin r\theta\}^m}{n\{x - \cos r\theta - \sqrt{(-1)}\sin r\theta\}} + \frac{\{\cos r\theta - \sqrt{(-1)}\sin r\theta\}^m}{n\{x - \cos r\theta + \sqrt{(-1)}\sin r\theta\}},$$

that is

$$\frac{\cos mr\theta + \sqrt{(-1)}\sin mr\theta}{n\{x - \cos r\theta - \sqrt{(-1)}\sin r\theta\}} + \frac{\cos mr\theta - \sqrt{(-1)}\sin mr\theta}{n\{x - \cos r\theta + \sqrt{(-1)}\sin r\theta\}},$$

that is $$\frac{2\cos mr\theta\,(x - \cos r\theta) - 2\sin mr\theta \sin r\theta}{n(x^2 - 2x\cos r\theta + 1)}.$$

Thus $$\frac{x^{m-1}}{x^n - 1} = \frac{1}{n(x-1)} + \frac{(-1)^m}{n(x+1)}$$

$$+ \frac{2}{n}\Sigma \frac{\cos mr\theta\,(x - \cos r\theta) - \sin mr\theta \sin r\theta}{(x - \cos r\theta)^2 + \sin^2 r\theta},$$

where Σ indicates a sum to be formed by giving to r all the even integral values from 2 to $n - 2$ inclusive. Hence

$$\int \frac{x^{m-1}dx}{x^n - 1} = \frac{1}{n}\log(x-1) + \frac{(-1)^m}{n}\log(x+1)$$

$$+ \frac{1}{n}\Sigma \cos mr\theta \log(x^2 - 2x\cos r\theta + 1) - \frac{2}{n}\Sigma \sin mr\theta \tan^{-1}\frac{x - \cos r\theta}{\sin r\theta}.$$

27. Required the integral of $\frac{x^{m-1}}{x^n - 1}$, n being supposed odd.

The real root of $x^n - 1 = 0$ is 1, and the imaginary roots are found from the expression $\cos r\theta \pm \sqrt{(-1)}\sin r\theta$, where $\theta = \frac{\pi}{n}$, and r takes in succession the values 2, 4, ... up to $n - 1$. Hence as before we shall find

$$\int \frac{x^{m-1}dx}{x^n - 1} = \frac{1}{n}\log(x-1) + \frac{1}{n}\Sigma \cos mr\theta \log(x^2 - 2x\cos r\theta + 1)$$

$$- \frac{2}{n}\Sigma \sin mr\theta \tan^{-1}\frac{x - \cos r\theta}{\sin r\theta}.$$

28. Required the integral of $\frac{x^{m-1}}{x^n+1}$, n being supposed even.

The equation $x^n+1=0$ has now no real root; the imaginary roots are found from the expression $\cos r\theta \pm \sqrt{(-1)} \sin r\theta$, where $\theta = \frac{\pi}{n}$, and r takes in succession the values 1, 3, ... up to $n-1$. And if a be a root of $x^n+1=0$, we have

$$\frac{\phi(a)}{F'(a)} = \frac{a^{m-1}}{na^{n-1}} = \frac{a^m}{na^n} = -\frac{a^m}{n};$$

thus the sum of the two fractions corresponding to a pair of imaginary roots is

$$-\frac{2}{n}\,\frac{\cos mr\theta\,(x-\cos r\theta) - \sin mr\theta \sin r\theta}{(x-\cos r\theta)^2 + \sin^2 r\theta}.$$

Hence

$$\int \frac{x^{m-1}\,dx}{x^n+1} = -\frac{1}{n}\Sigma \cos mr\theta \log(x^2 - 2x\cos r\theta + 1)$$

$$+\frac{2}{n}\Sigma \sin mr\theta \tan^{-1}\frac{x-\cos r\theta}{\sin r\theta},$$

where Σ indicates a sum to be formed by giving to r all the odd integral values from 1 to $n-1$ inclusive.

29. Required the integral of $\frac{x^{m-1}}{x^n+1}$, n being supposed odd.

The real root of $x^n+1=0$ is in this case -1, and the imaginary roots are found from the expression $\cos r\theta \pm \sqrt{(-1)} \sin r\theta$, where $\theta = \frac{\pi}{n}$ and r takes in succession the values 1, 3, ... up to $n-2$. Hence we shall obtain

$$\int \frac{x^{m-1}\,dx}{x^n+1} = \frac{(-1)^{m-1}}{n}\log(x+1)$$

$$-\frac{1}{n}\Sigma \cos mr\theta \log(x^2 - 2x\cos r\theta + 1) + \frac{2}{n}\Sigma \sin mr\theta \tan^{-1}\frac{x-\cos r\theta}{\sin r\theta}.$$

EXAMPLES.

1. $\int \frac{dx}{x^3-1} = \frac{1}{6} \log \frac{(x-1)^2}{x^2+x+1} - \frac{1}{\sqrt{3}} \tan^{-1} \frac{2x+1}{\sqrt{3}}$.

2. $\int \frac{x^2-1}{x^2-4} dx = x + \log \left(\frac{x-2}{x+2}\right)^{\frac{3}{4}}$.

3. $\int \frac{x^3 dx}{x^2+7x+12} = \frac{x^2}{2} - 7x + 64 \log (x+4) - 27 \log (x+3)$.

4. $\int \frac{dx}{a^4-x^4} = \frac{1}{2a^3} \tan^{-1} \frac{x}{a} + \frac{1}{4a^3} \log \frac{a+x}{a-x}$.

5. $\int \frac{2x^2-3a^2}{x^4-a^4} dx = \frac{5}{2a} \tan^{-1} \frac{x}{a} - \frac{1}{4a} \log \frac{x-a}{x+a}$.

6. $\int \frac{dx}{(x^2+1)(x^2+x+1)} = \frac{1}{2} \log \frac{x^2+x+1}{x^2+1} + \frac{1}{\sqrt{3}} \tan^{-1} \frac{2x+1}{\sqrt{3}}$.

7. $\int \frac{x^2 dx}{x^4+x^2-2} = \frac{1}{6} \log \frac{x-1}{x+1} + \frac{\sqrt{2}}{3} \tan^{-1} \frac{x}{\sqrt{2}}$.

8. $\int \frac{x^2-1}{x^4+x^2+1} dx = \frac{1}{2} \log \frac{x^2-x+1}{x^2+x+1}$.

9. $\int \frac{x^2-3x+3}{(x-1)(x-2)} dx = x + \log \frac{x-2}{x-1}$.

10. $\int \frac{(3x-1)\,dx}{x^3-x^2-2x} = \frac{1}{2} \log x + \frac{5}{6} \log (x-2) - \frac{4}{3} \log (x+1)$.

11. $\int \frac{dx}{(x^2+a^2)(x+b)} = \frac{1}{b^2+a^2} \left\{\log \frac{x+b}{\sqrt{(x^2+a^2)}} + \frac{b}{a} \tan^{-1} \frac{x}{a}\right\}$.

12. $\int \frac{dx}{x(1+x+x^2+x^3)} = \log x - \frac{1}{2} \log (1+x) - \frac{1}{4} \log (1+x^2)$
$- \frac{1}{2} \tan^{-1} x$.

13. $$\int \frac{dx}{(x-1)^2(x^2+1)^2} = -\frac{1}{4(x-1)} - \tfrac{1}{2}\log(x-1)$$
$$+ \tfrac{1}{4}\tan^{-1}x - \frac{1}{4(x^2+1)} + \tfrac{1}{4}\log(x^2+1).$$

14. $$\int \frac{xdx}{(1+x)(1+2x)^2(1+x^2)} = \frac{2}{5}\frac{1}{1+2x} - \frac{1}{2}\log(1+x)$$
$$-\frac{7}{100}\log(1+x^2) + \frac{16}{25}\log(1+2x) + \frac{1}{50}\tan^{-1}x.$$

15. $$\int \frac{x^2dx}{x^4+1} = \frac{1}{4\sqrt{2}}\log\frac{x^2-x\sqrt{2}+1}{x^2+x\sqrt{2}+1}$$
$$+\frac{1}{2\sqrt{2}}\{\tan^{-1}(x\sqrt{2}+1)+\tan^{-1}(x\sqrt{2}-1)\}$$

16. $$\int \frac{x^3dx}{x^6+1} = \frac{1}{12}\log(x^4-x^2+1) - \frac{1}{6}\log(x^2+1)$$
$$+\frac{1}{2\sqrt{3}}\{\tan^{-1}(2x-\sqrt{3})-\tan^{-1}(2x+\sqrt{3})\}.$$

17. $\int \frac{dy}{\sqrt[3]{(1-y^3)}}$. Assume $1-y^3=y^3z^3$.

18. $\int \frac{dx}{(1+x)\sqrt[3]{(1+3x+3x^2)}}$. Assume $y=\frac{x}{1+x}$.

CHAPTER III.

FORMULÆ OF REDUCTION.

30. LET $a + bx^n$ be denoted by X; by integration by parts we have

$$\int x^{m-1} X^p dx = \frac{X^p x^m}{m} - \int \frac{x^m}{m} p X^{p-1} \frac{dX}{dx} dx$$

$$= \frac{X^p x^m}{m} - \frac{bnp}{m} \int x^{m+n-1} X^{p-1} dx \text{..........} (1).$$

The equation (1) is called a *formula of reduction;* by means of it we make the integral of $x^{m-1}X^p$ depend on that of $x^{m+n-1}X^{p-1}$. In the same way the latter integral can be made to depend on that of $x^{m+2n-1}X^{p-2}$; and thus, if p be an integer we may proceed until we arrive at $x^{m+np-1}X^{p-p}$, that is x^{m+np-1}, which is immediately integrable.

From (1), by transposition,

$$\int x^{m+n-1} X^{p-1} dx = \frac{x^m X^p}{bnp} - \frac{m}{bnp} \int x^{m-1} X^p dx.$$

Change m into $m-n$ and p into $p+1$; thus

$$\int x^{m-1} X^p dx = \frac{x^{m-n} X^{p+1}}{bn(p+1)} - \frac{m-n}{bn(p+1)} \int x^{m-n-1} X^{p+1} dx \text{...} (2).$$

This formula may be used when we wish to make the integral of $x^m X^p$ depend upon another in which the exponent of x is diminished and that of X increased. For example, if $m = 3$, $n = 2$, and $p = -\frac{3}{2}$, we have

$$\int \frac{x^2 dx}{(a+bx^2)^{\frac{3}{2}}} = -\frac{x}{b\sqrt{(a+bx^2)}} + \frac{1}{b} \int \frac{dx}{\sqrt{(a+bx^2)}}.$$

The latter integral has already been determined, and thus the proposed integration is accomplished.

Since $\int x^{m-1}X^p dx = \int x^{m-1}X^{p-1}(a+bx^n)\,dx$

$$= a\int x^{m-1}X^{p-1}dx + b\int x^{m+n-1}X^{p-1}dx,$$

we have by (1)

$$\frac{x^m X^p}{m} - \frac{bnp}{m}\int x^{m+n-1}X^{p-1}\,dx = a\int x^{m-1}X^{p-1}\,dx + b\int x^{m+n-1}X^{p-1}dx,$$

therefore $\int x^{m-1}X^{p-1}\,dx = \frac{x^m X^p}{am} - \frac{b(m+np)}{am}\int x^{m+n-1}X^{p-1}\,dx.$

Change p into $p+1$, and we have

$$\int x^{m-1}X^p dx = \frac{x^m X^{p+1}}{am} - \frac{b(m+np+n)}{am}\int x^{m+n-1}X^p dx \text{......(3)}.$$

Change m into $m-n$ and transpose, then

$$\int x^{m-1}X^p\,dx = \frac{x^{m-n}X^{p+1}}{b(m+np)} - \frac{(m-n)a}{b(m+np)}\int x^{m-n-1}X^p dx \text{.......(4)}.$$

We have already obtained from (1) by transposition

$$\int x^{m+n-1}X^{p-1}dx = \frac{x^m X^p}{bnp} - \frac{m}{bnp}\int x^{m-1}X^p\,dx;$$

also $\int x^{m-1}X^p dx = a\int x^{m-1}X^{p-1}dx + b\int x^{m+n-1}X^{p-1}dx;$

therefore $\int x^{m-1}X^p\,dx = a\int x^{m-1}X^{p-1}dx + \frac{x^m X^p}{np} - \frac{m}{np}\int x^{m-1}X^p dx;$

therefore $\int x^{m-1}X^p dx = \frac{x^m X^p}{m+np} + \frac{anp}{m+np}\int x^{m-1}X^{p-1}dx \text{.......(5)}.$

Change p into $p+1$ and transpose; thus

$$\int x^{m-1}X^p dx = -\frac{x^m X^{p+1}}{an(p+1)} + \frac{m+np+n}{an(p+1)}\int x^{m-1}X^{p+1}dx \text{.....(6)}.$$

31. If an example is proposed to which one of the preceding formulæ is applicable, we may either quote that particular formula or may obtain the required result independently. Thus, suppose we require $\int \frac{x^m dx}{\sqrt{(c^2 - x^2)}}$; we have

$$\int \frac{x^m dx}{\sqrt{(c^2 - x^2)}} = -\int \frac{d\sqrt{(c^2 - x^2)}}{dx} x^{m-1} dx$$

$$= -\sqrt{(c^2 - x^2)}\, x^{m-1} + (m-1)\int x^{m-2}\sqrt{(c^2 - x^2)}\, dx$$

$$= -\sqrt{(c^2 - x^2)}\, x^{m-1} + (m-1)\int \frac{(c^2 - x^2)\, x^{m-2}\, dx}{\sqrt{(c^2 - x^2)}}.$$

By transposition,

$$(1 + m - 1)\int \frac{x^m dx}{\sqrt{(c^2 - x^2)}} = -\sqrt{(c^2 - x^2)}\, x^{m-1} + (m-1)\, c^2 \int \frac{x^{m-2} dx}{\sqrt{(c^2 - x^2)}},$$

therefore

$$\int \frac{x^m dx}{\sqrt{(c^2 - x^2)}} = -\frac{x^{m-1}\sqrt{(c^2 - x^2)}}{m} + \frac{(m-1)\, c^2}{m}\int \frac{x^{m-2} dx}{\sqrt{(c^2 - x^2)}} \ \ldots\ldots (1).$$

This result agrees with the equation (4) of the preceding article if we make $a = c^2$, $b = -1$, $n = 2$, $p = -\frac{1}{2}$, and change m into $m + 1$.

Again, suppose we require $\int \frac{dx}{x^m \sqrt{(a^2 + x^2)}}$. We have

$$\int \frac{dx}{x^m \sqrt{(a^2 + x^2)}} = \int \frac{d\sqrt{(a^2 + x^2)}}{dx} \frac{1}{x^{m+1}} dx$$

$$= \frac{\sqrt{(a^2 + x^2)}}{x^{m+1}} + (m+1)\int \frac{\sqrt{(a^2 + x^2)}}{x^{m+2}} dx$$

$$= \frac{\sqrt{(a^2 + x^2)}}{x^{m+1}} + (m+1)\int \frac{a^2 + x^2}{x^{m+2}\sqrt{(a^2 + x^2)}} dx.$$

By transposition,

$$(m+1)\, a^2 \int \frac{dx}{x^{m+2}\sqrt{(a^2 + x^2)}} = -\frac{\sqrt{(a^2 + x^2)}}{x^{m+1}} - m\int \frac{dx}{x^m \sqrt{(a^2 + x^2)}},$$

and by changing m into $m-2$ we obtain

$$\int \frac{dx}{x^m \sqrt{(a^2+x^2)}} = -\frac{\sqrt{(a^2+x^2)}}{(m-1)\,a^2 x^{m-1}} - \frac{m-2}{(m-1)\,a^2}\int \frac{dx}{x^{m-2}\sqrt{(a^2+x^2)}} \quad \ldots\ldots\ldots\ldots(2).$$

Another example is furnished by $\int \frac{x^m dx}{\sqrt{(2ax-x^2)}}$, which may be written $\int \frac{x^{m-\frac{1}{2}}dx}{\sqrt{(2a-x)}}$; if in equation (4) of the preceding article we make $b=-1$, $n=1$, $p=-\frac{1}{2}$, and change a and m into $2a$ and $m+\frac{1}{2}$ respectively, we have

$$\int \frac{x^m dx}{\sqrt{(2ax-x^2)}} = -\frac{x^{m-1}\sqrt{(2ax-x^2)}}{m} + \frac{a(2m-1)}{m}\int \frac{x^{m-1}dx}{\sqrt{(2ax-x^2)}} \quad \ldots\ldots\ldots\ldots(3),$$

which of course may be found independently.

32. In equation (6) of Art. 30 put $a=c^2$, $m=1$, $n=2$, $b=1$, and $p=-r$; thus

$$\int \frac{dx}{(x^2+c^2)^r} = \frac{x}{2(r-1)\,c^2(x^2+c^2)^{r-1}} + \frac{2r-3}{2(r-1)\,c^2}\int \frac{dx}{(x^2+c^2)^{r-1}}.$$

This formula will serve to reduce the form

$$\int \frac{(Ax+B)\,dx}{(x^2-2\alpha x+\alpha^2+\beta^2)^r},$$

which occurs in Art. 18; for this last expression may be written thus,

$$\int \frac{A(x-\alpha)\,dx}{\{(x-\alpha)^2+\beta^2\}^r} + (A\alpha+B)\int \frac{dx}{\{(x-\alpha)^2+\beta^2\}^r},$$

that is

$$-\frac{A}{2(r-1)}\frac{1}{\{(x-\alpha)^2+\beta^2\}^{r-1}} + (A\alpha+B)\int \frac{dx}{\{(x-\alpha)^2+\beta^2\}^r}.$$

By putting $x-\alpha=x'$, we have

$$\int \frac{dx}{\{(x-\alpha)^2+\beta^2\}^r} = \int \frac{dx'}{\{x'^2+\beta^2\}^r},$$

and thus the above formula becomes applicable.

33. These formulæ of reduction are most useful when the integral has to be taken between certain limits. Suppose $\phi(x)$, $\chi(x)$, $\psi(x)$, functions of x, such that

$$\int \phi(x)\,dx = \chi(x) + \int \psi(x)\,dx,$$

then
$$\int_a^b \phi(x)dx = \chi(b) - \chi(a) + \int_a^b \psi(x)\,dx,$$

as is obvious from Art. 3.

For example, it may be shewn that

$$\int (c^2 - x^2)^{\frac{n}{2}}\,dx = \frac{x(c^2 - x^2)^{\frac{n}{2}}}{n+1} + \frac{nc^2}{n+1}\int (c^2 - x^2)^{\frac{n}{2}-1}\,dx;$$

suppose $\frac{n}{2}$ a *positive* quantity, then $x(c^2 - x^2)^{\frac{n}{2}}$ vanishes both when $x = 0$ and when $x = c$. Hence

$$\int_0^c (c^2 - x^2)^{\frac{n}{2}}\,dx = \frac{nc^2}{n+1}\int_0^c (c^2 - x^2)^{\frac{n}{2}-1}\,dx.$$

The following is a similar example. By integration by parts

$$\int x^{r-1}(1-x)^{n-1}\,dx = -\frac{(1-x)^n}{n}x^{r-1} + \frac{r-1}{n}\int x^{r-2}(1-x)^n\,dx.$$

Hence
$$\int_0^1 x^{r-1}(1-x)^{n-1}\,dx = \frac{r-1}{n}\int_0^1 x^{r-2}(1-x)^n\,dx.$$

Thus if r be an integer we may reduce the integral to $\int_0^1 (1-x)^{n+r-2}\,dx$, that is to $\frac{1}{n+r-1}$; hence

$$\int_0^1 x^{r-1}(1-x)^{n-1}\,dx = \frac{(r-1)(r-2)\ldots\ldots 3.2.1}{n(n+1)(n+2)\ldots\ldots(n+r-1)}.$$

34. The integration of trigonometrical functions is facilitated by formulæ of reduction. Let $\phi(\sin x, \cos x)$ denote

any function of $\sin x$ and $\cos x$; then if we put $\sin x = z$, we have

$$\int \phi(\sin x, \cos x)\, dx = \int \phi\{z, \sqrt{(1-z^2)}\} \frac{dx}{dz}\, dz$$

$$= \int \phi\{z, \sqrt{(1-z^2)}\} \frac{dz}{\sqrt{(1-z^2)}} \quad \text{........(1).}$$

For example, let $\phi(\sin x, \cos x) = \sin^p x \cos^q x$; then

$$\int \sin^p x \cos^q x\, dx = \int z^p (1-z^2)^{\frac{1}{2}(q-1)} dz \text{.........(2).}$$

If in the six formulæ of Art. 30 we put $a = 1$, $b = -1$, $n = 2$, $p = \frac{1}{2}(q-1)$, we have

$$\int z^{m-1} (1-z^2)^{\frac{1}{2}(q-1)} dz$$

$$= \frac{z^m (1-z^2)^{\frac{1}{2}(q-1)}}{m} + \frac{q-1}{m} \int z^{m+1} (1-z^2)^{\frac{1}{2}(q-3)} dz$$

$$= -\frac{z^{m-2} (1-z^2)^{\frac{1}{2}(q+1)}}{q+1} + \frac{m-2}{q+1} \int z^{m-3} (1-z^2)^{\frac{1}{2}(q+1)} dz$$

$$= \frac{z^m (1-z^2)^{\frac{1}{2}(q+1)}}{m} + \frac{m+q+1}{m} \int z^{m+1} (1-z^2)^{\frac{1}{2}(q-1)} dz$$

$$= -\frac{z^{m-2} (1-z^2)^{\frac{1}{2}(q+1)}}{m+q-1} + \frac{m-2}{m+q-1} \int z^{m-3} (1-z^2)^{\frac{1}{2}(q-1)} dz$$

$$= \frac{z^m (1-z^2)^{\frac{1}{2}(q-1)}}{m+q-1} + \frac{q-1}{m+q-1} \int z^{m-1} (1-z^2)^{\frac{1}{2}(q-3)} dz$$

$$= -\frac{z^m (1-z^2)^{\frac{1}{2}(q+1)}}{q+1} + \frac{m+q+1}{q+1} \int z^{m-1} (1-z^2)^{\frac{1}{2}(q+1)} dz.$$

If we put $m = p + 1$, and $z = \sin x$, the first of the above equations becomes

$$\int \sin^p x \cos^q x\, dx = \frac{\sin^{p+1} x \cos^{q-1} x}{p+1} + \frac{q-1}{p+1} \int \sin^{p+2} x \cos^{q-2} x\, dx,$$

and similarly the other five equations may be expressed.

35. The following is a very important case:

$$\int \sin^n x\,dx = -\int \frac{d\cos x}{dx} \sin^{n-1} x\,dx$$

$$= -\cos x \sin^{n-1} x + (n-1) \int \cos^2 x \sin^{n-2} x\,dx$$

$$= -\cos x \sin^{n-1} x + (n-1) \int (1 - \sin^2 x) \sin^{n-2} x\,dx.$$

Transposing, we have

$$n \int \sin^n x\,dx = -\cos x \sin^{n-1} x + (n-1) \int \sin^{n-2} x\,dx;$$

therefore $\displaystyle\int \sin^n x\,dx = -\frac{\cos x \sin^{n-1} x}{n} + \frac{n-1}{n} \int \sin^{n-2} x\,dx.$

From the last equation we deduce

$$\int_0^{\frac{1}{2}\pi} \sin^n x\,dx = \frac{n-1}{n} \int_0^{\frac{1}{2}\pi} \sin^{n-2} x\,dx.$$

Similarly $\displaystyle\int_0^{\frac{1}{2}\pi} \sin^{n-2} x\,dx = \frac{n-3}{n-2} \int_0^{\frac{1}{2}\pi} \sin^{n-4} x\,dx.$

Proceeding thus, if n be an *even* integer we shall arrive at $\int_0^{\frac{1}{2}\pi} dx$ or $\frac{1}{2}\pi$; if n be an *odd* integer we shall arrive at $\int_0^{\frac{1}{2}\pi} \sin x\,dx$, which is unity. Hence, if n be an integer,

$$\int_0^{\frac{1}{2}\pi} \sin^n x\,dx = \frac{(n-1)(n-3)(n-5)\ldots\ldots 1}{n(n-2)(n-4)\ldots\ldots 2} \frac{\pi}{2} \quad (n \text{ even}),$$

$$\int_0^{\frac{1}{2}\pi} \sin^n x\,dx = \frac{(n-1)(n-3)(n-5)\ldots\ldots 2}{n(n-2)(n-4)\ldots\ldots 3} \quad (n \text{ odd}).$$

These two results hold if we change $\sin x$ into $\cos x$, as will be found on investigation.

36. From the preceding results we may deduce an important theorem, called Wallis's Formula.

Suppose n even; then

$$\int_0^{\frac{1}{2}\pi} \sin^n x\,dx = \frac{n-1}{n} \cdot \frac{n-3}{n-2} \cdot \frac{n-5}{n-4} \cdots\cdots \frac{3}{4} \cdot \frac{1}{2} \cdot \frac{\pi}{2} \cdots\cdots (1),$$

$$\int_0^{\frac{1}{2}\pi} \sin^{n-1} x\,dx = \frac{n-2}{n-1} \cdot \frac{n-4}{n-3} \cdot \frac{n-6}{n-5} \cdots\cdots \frac{2}{3} \cdots\cdots\cdots (2).$$

Now it is obvious that $\int_0^{\frac{1}{2}\pi} \sin^{n-1} x\,dx$ is less than $\int_0^{\frac{1}{2}\pi} \sin^{n-2} x\,dx$ and greater than $\int_0^{\frac{1}{2}\pi} \sin^n x\,dx$; because each element of the first integral is less than the corresponding element of the second integral and greater than the corresponding element of the third integral. And it has been shewn that

$$\frac{\int_0^{\frac{1}{2}\pi} \sin^n x\,dx}{\int_0^{\frac{1}{2}\pi} \sin^{n-2} x\,dx} = \frac{n-1}{n}.$$

Therefore $\dfrac{\int_0^{\frac{1}{2}\pi} \sin^n x\,dx}{\int_0^{\frac{1}{2}\pi} \sin^{n-1} x\,dx}$ is less than 1 and greater than $\dfrac{n-1}{n}$.

Hence the ratio of the right-hand member of (1) to the right-hand member of (2) is less than unity and greater than $\dfrac{n-1}{n}$; thus

$$\frac{\pi}{2} > \frac{2.2.4.4.6.6\ldots\ldots(n-2)(n-2)}{1.3.3.5.5.7\ldots\ldots(n-3)(n-1)},$$

$$\text{and} < \frac{2.2.4.4.6.6\ldots\ldots(n-2)(n-2)}{1.3.3.5.5.7\ldots\ldots(n-3)(n-1)} \frac{n}{n-1}.$$

EXAMPLES.

1. $$\int (a^2+x^2)^{\frac{n}{2}} dx = \frac{x(a^2+x^2)^{\frac{n}{2}}}{n+1} + \frac{na^2}{n+1} \int (a^2+x^2)^{\frac{n}{2}-1} dx.$$

2. $$\int x^m \sqrt{(2ax - x^2)}\, dx = -\frac{x^{m-1}(2ax - x^2)^{\frac{3}{2}}}{m+2} + \frac{a(2m+1)}{m+2}\int x^{m-1}\sqrt{(2ax - x^2)}\, dx.$$

3. $$\int x\sqrt{(2ax - x^2)}\, dx = -\tfrac{1}{3}(2ax - x^2)^{\frac{3}{2}} + a\int \sqrt{(2ax - x^2)}\, dx.$$

4. $$\int_0^{2a} x\sqrt{(2ax - x^2)}\, dx = \frac{\pi a^3}{2}.$$

5. $$\int x^2\sqrt{(2ax - x^2)}\, dx = -\frac{x}{4}(2ax - x^2)^{\frac{3}{2}} + \frac{5a}{4}\int x\sqrt{(2ax - x^2)}\, dx.$$

6. $$\int_0^{2a} x^2\sqrt{(2ax - x^2)}\, dx = \frac{5a^4\pi}{8}.$$

7. $$\int_0^{2a} x^3\sqrt{(2ax - x^2)}\, dx = \frac{7\pi a^5}{8}.$$

8. $$\int x^n (\log x)^m\, dx = \frac{x^{n+1}(\log x)^m}{n+1} - \frac{m}{n+1}\int x^n (\log x)^{m-1}\, dx.$$

9. $$\int x^n (\log x)^2\, dx = \frac{x^{n+1}}{n+1}\left\{(\log x)^2 - \frac{2}{n+1}\log x + \frac{2}{(n+1)^2}\right\}.$$

10. $$\int_0^{\frac{1}{4}\pi} \sec^4\theta\, d\theta = \tfrac{4}{3}.$$

11. $$\int_0^a \frac{x^2\sqrt{(a-x)}}{\sqrt{(a+x)}}\, dx = \left(\frac{\pi}{4} - \frac{2}{3}\right)a^3.$$

12. $$\int \sin^3\theta\cos^3\theta\, d\theta = -\tfrac{1}{4}\cos^4\theta + \tfrac{1}{6}\cos^6\theta.$$

13. $$\int \frac{d\theta}{\sin^4\theta\cos^4\theta} = 3(\tan\theta - \cot\theta) + \tfrac{1}{3}(\tan^3\theta - \cot^3\theta).$$

14. $$\int \frac{\sin^2\theta\, d\theta}{\cos^3\theta} = \frac{\sin\theta}{2\cos^2\theta} + \tfrac{1}{4}\log\frac{1 - \sin\theta}{1 + \sin\theta}.$$

15. $\int_{-\frac{1}{4}\pi}^{\frac{1}{4}\pi} (\cos 2\theta)^{\frac{3}{2}} \cos\theta \, d\theta = \frac{3\pi\sqrt{(2)}}{16}$.

Assume $\sqrt{(2)} \sin\theta = \sin\phi$.

16. $\int_0^a \sqrt{(a^2 - x^2)} \cos^{-1}\frac{x}{a} \, dx = \left(1 + \frac{\pi^2}{4}\right)\frac{a^2}{4}$.

17. $\int_0^{2a} \left(\operatorname{vers}^{-1}\frac{x}{a}\right)^2 dx = (\pi^2 - 4)\, a$.

18. $\int_0^{\frac{1}{2}\pi} \frac{\sin^3 x \, dx}{1 + c\cos x} = \frac{c^2 - 1}{c^3} \log(1 + c) + \frac{2 - c}{2c^2}$.

19. If $\phi(n) = \int (1 + c\cos x)^{-n} dx$, shew that

$$(n-1)(1 - c^2)\,\phi(n) = -c\sin x\,(1 + c\cos x)^{-n+1} + (2n-3)\,\phi(n-1) - (n-2)\,\phi(n-2).$$

20. $\int_0^{2a} \sqrt{(2ax - x^2)} \operatorname{vers}^{-1}\frac{x}{a} \, dx = \frac{\pi^2 a^2}{4}$.

21. $\int_0^{2a} x\sqrt{(2ax - x^2)} \operatorname{vers}^{-1}\frac{x}{a} \, dx = \frac{4a^3}{9} + \frac{\pi^2 a^3}{4}$.

22. $\int_0^{\frac{1}{4}\pi} (\tan x)^7 \, dx = \frac{5}{12} - \frac{1}{2}\log 2$.

23. $\int_0^{\frac{1}{2}\pi} \frac{dx}{\sqrt{(1 - c^2\sin^2 x)}} = \frac{\pi}{2}\left\{1 + (\tfrac{1}{2})^2 c^2 + \left(\frac{1.3}{2.4}\right)^2 c^4 + \left(\frac{1.3.5}{2.4.6}\right)^2 c^6 + \ldots\right\}$

c being < 1.

24. Let $P = Ax^a + Bx^b + Cx^c + \ldots$, $V_{m,n} = \int x^m P^n \, dx$,

$\alpha = m + 1 + na$, $\beta = m + 1 + nb$, $\gamma = m + 1 + nc \ldots$

Then

$$V_{m,n} = AV_{m+a,\,n-1} + BV_{m+b,\,n-1} + CV_{m+c,\,n-1} + \ldots$$

$$x^{m+1}P^n = \alpha AV_{m+a,\,n-1} + \beta BV_{m+b,\,n-1} + \gamma CV_{m+c,\,n-1} + \ldots$$

(*Cambridge and Dublin Mathematical Journal*, Vol. III. p. 242.)

CHAPTER IV.

MISCELLANEOUS REMARKS.

37. We have at the beginning of this book defined the *integral* of $\phi(x)$ between assigned limits a and b as the limit of a certain sum $\Sigma\phi(x)\,\Delta x$, and have denoted this limit by $\int_a^b \phi(x)\,dx$. We have shewn that this limit is known as soon as we know the function $\psi(x)$ of which $\phi(x)$ is the differential coefficient. In the pages immediately following we gave methods for finding $\psi(x)$ in different cases. We shall now add some miscellaneous remarks and theorems, some of which will recall the attention of the student to the process of summation which we placed at the foundation of the subject.

38. Suppose we wish to find the integral of $\sin x$ between limits a and b *immediately from the definition*. By Art. 4 we have to find the limit when n is infinite of

$$h\,[\sin a + \sin(a+h) + \sin(a+2h) \ldots\ldots + \sin\{a+(n-1)\,h\}],$$

where $h = \frac{1}{n}(b-a)$.

It is known from Trigonometry that this series

$$= \frac{h \sin\left(a + \frac{n-1}{2}h\right)\sin\frac{nh}{2}}{\sin\frac{h}{2}} = \frac{h \sin\left(a + \frac{b-a}{2} - \frac{h}{2}\right)\sin\frac{b-a}{2}}{\sin\frac{h}{2}}.$$

The limit of $\dfrac{h}{\sin\dfrac{h}{2}}$ when n is infinite and therefore h zero is 2; hence the required integral is

$$2\sin\frac{b+a}{2}\sin\frac{b-a}{2}=\cos a-\cos b.$$

39. Required the limit when n is made infinite of the series

$$\frac{n}{n^2}+\frac{n}{1+n^2}+\frac{n}{2^2+n^2}+\frac{n}{3^2+n^2}\ \ldots\ldots+\frac{n}{(n-1)^2+n^2}.$$

This series may be written

$$\frac{1}{n}\left\{\frac{1}{1}+\frac{1}{1+\left(\frac{1}{n}\right)^2}+\frac{1}{1+\left(\frac{2}{n}\right)^2}+\frac{1}{1+\left(\frac{3}{n}\right)^2}\ \ldots\ldots+\frac{1}{1+\left(\frac{n-1}{n}\right)^2}\right\};$$

putting h for $\frac{1}{n}$, we obtain

$$h\left\{\frac{1}{1}+\frac{1}{1+h^2}+\frac{1}{1+(2h)^2}\ \ldots\ldots+\frac{1}{1+(n-1)^2h^2}\right\}.$$

Comparing this with Art. 4 we see that the required limit is what we denote by $\int_0^1\frac{dx}{1+x^2}$. Now $\int\frac{dx}{1+x^2}=\tan^{-1}x$; hence $\frac{\pi}{4}$ is the required limit.

40. We define $\int_a^b\phi(x)\,dx$ as the limit when n is infinite of

$$h_1\phi(a)+h_2\phi(x_1)\ \ldots\ldots+h_n\phi(x_{n-1}).$$

Now let A and B be the greatest and least values which $\phi(x)$ takes between the limits a and b; then the series is less than

$$(h_1+h_2+\ldots\ldots+h_n)\,A,$$

and is greater than

$$(h_1 + h_2 + \ldots\ldots + h_n) B;$$

that is, the series lies between

$$(b-a) A \text{ and } (b-a) B.$$

The limit must therefore be equal to $(b-a) C$, where C is some quantity lying between A and B; but since $\phi(x)$ is supposed continuous, it must, while x ranges from a to b, pass through every value between A and B, and must therefore be equal to C when x has some value between a and b. Thus $C = \phi\{a + \theta(b-a)\}$, where θ is some proper fraction, and

$$\int_a^b \phi(x)\, dx = (b-a)\, \phi\{a + \theta(b-a)\}.$$

Similarly if $\psi(x)$ retains the same sign while x lies between a and b, we may prove that

$$\int_a^b \phi(x)\, \psi(x)\, dx = \phi\{a + \theta(b-a)\} \int_a^b \psi(x)\, dx.$$

41. The truth of the equation

$$\int_a^b \phi(x)\, dx = \int_a^c \phi(x)\, dx + \int_c^b \phi(x)\, dx \ldots\ldots\ldots\ldots (1)$$

will appear immediately; for suppose $\psi(x)$ to be the integral of $\phi(x)$, then we have on the left-hand side

$$\psi(b) - \psi(a),$$

and on the right hand

$$\psi(c) - \psi(a) + \psi(b) - \psi(c).$$

In like manner the equation

$$\int_a^b \phi(x)\, dx = -\int_b^a \phi(x)\, dx \ldots\ldots\ldots\ldots (2)$$

is obviously true. We may shew also that

$$\int_0^a \phi(x)\, dx = \int_0^a \phi(a-x)\, dx \ldots\ldots\ldots\ldots (3).$$

For putting $a - x = z$ we have

$$\int \phi (a - x)\, dx = - \int \phi (z)\, dz,$$

therefore $$\int_0^a \phi (a - x)\, dx = - \int_a^0 \phi (z)\, dz$$

$$= \int_0^a \phi (z)\, dz, \text{ by (2).}$$

Of course $\int_0^a \phi (z)\, dz = \int_0^a \phi (x)\, dx$, since it is indifferent whether we use the symbol x or z in obtaining a result which does not involve x or z.

We have from (1)

$$\int_0^{2a} \phi (x)\, dx = \int_0^a \phi (x)\, dx + \int_a^{2a} \phi (x)\, dx.$$

The second integral, by changing x into $2a - x'$, will be found equal to

$$\int_0^a \phi (2a - x')\, dx' \text{ or } \int_0^a \phi (2a - x)\, dx.$$

Hence

$$\int_0^{2a} \phi (x)\, dx = \int_0^a \{\phi (x) + \phi (2a - x)\}\, dx.$$

Hence, if $\phi (x) = \phi (2a - x)$ for all values of x comprised between 0 and a, we have

$$\int_0^{2a} \phi (x)\, dx = 2 \int_0^a \phi (x)\, dx \quad \text{.................} \quad (4),$$

and if $\phi (2a - x) = - \phi (x)$, we have

$$\int_0^{2a} \phi (x)\, dx = 0 \quad \text{.........................} \quad (5).$$

For example,

$$\int_0^{\pi} \sin^3 \theta\, d\theta = 2 \int_0^{\frac{1}{2}\pi} \sin^3 \theta\, d\theta \text{ ...... by (4),}$$

and $$\int_0^\pi \cos^3\theta\,d\theta = 0 \text{ ...... by (5).}$$

42. Such equations as those just given should receive careful attention from the student, and he should not leave them until he recognizes their obvious and self-evident truth. $\int_0^\pi \cos^3\theta\,d\theta$ is by definition the limit when n is infinite of the series

$$h\{\cos^3 h + \cos^3 2h + \cos^3 3h \text{ ......} + \cos^3 (n-1)\,h\},$$

where $nh = \pi$. Now

$$\cos^3 h = -\cos^3 (n-1)\,h, \quad \cos^3 2h = -\cos^3 (n-2)\,h \text{......};$$

thus the positive terms of the series just balance the negative terms and leave zero as the result.

In the same way the truth of $\int_0^\pi \sin^3\theta\,d\theta = 2\int_0^{\frac{1}{2}\pi} \sin^3\theta\,d\theta$ follows *immediately* from the definition of integration, and the fact that the sine of an angle is equal to the sine of the supplemental angle.

43. Suppose b greater than a and $\phi(x)$ always positive between the limits a and b of x; then every term in the series $\Sigma\phi(x)\,\Delta x$ is positive, and hence the limit $\int_a^b \phi(x)\,dx$ *must be a positive quantity.*

44. All the statements which have been made suppose that the function which is to be integrated is always finite between the limits of integration; for it must be remembered that this condition was expressly introduced in the fundamental proposition, Art. 2. If therefore the function to be integrated becomes infinite between the limits of integration, the rules of integration cannot be applied; at least the case must be specially examined.

45. Consider $\int_0^a \frac{dx}{\sqrt{(1-x)}}$; the value of this integral is $2 - 2\sqrt{(1-a)}$. Here the function to be integrated becomes infinite when $x = 1$; but the expression $2 - 2\sqrt{(1-a)}$ is finite when $a = 1$. Hence in this case we may write

$\int_0^1 \frac{dx}{\sqrt{(1-x)}} = 2$, provided that we regard this as an abbreviation of the following statement: "$\int_0^a \frac{dx}{\sqrt{(1-x)}}$ is always finite if a be any quantity less than unity, and by taking a sufficiently near to unity, we can make the value of the integral differ as little as we please from 2."

46. Next take $\int_0^a \frac{dx}{1-x}$; the value of this integral is $-\log(1-a)$, which increases indefinitely as a approaches to unity. Hence in this case we may write $\int_0^1 \frac{dx}{1-x} = \infty$ provided that we regard this as an abbreviation of the following statement: "$\int_0^a \frac{dx}{1-x}$ increases indefinitely as a approaches to unity, and by taking a sufficiently near to unity we can make the integral greater than any assigned quantity."

47. Next consider $\int \frac{dx}{(1-x)^2}$; the integral here is $\frac{1}{1-x}$. If without remarking that the function to be integrated becomes infinite when $x = 1$, we propose to find the value of the integral between the limits 0 and 2, we obtain $-1-1$, that is -2. But this is obviously false, for in this case every term of the series indicated by $\Sigma\phi(x)\Delta x$ is positive, and therefore the limit cannot be negative. In fact $\int_0^1 \frac{dx}{(1-x)^2}$ and $\int_1^2 \frac{dx}{(1-x)^2}$ are both infinite. This example shews that the ordinary rules for integrating between assigned limits cannot be used when the function to be integrated becomes infinite between those limits.

48. In the fundamental investigation in Art. 2, of the value of $\int_a^b \phi(x)\,dx$, the limits a and b are supposed to be *finite* as well as the function $\phi(x)$. But we shall often find it convenient to suppose one or both of the limits *infinite*, as we will now indicate by examples.

Consider $\int \frac{dx}{1+x^2}$; the integral is $\tan^{-1}x$. Hence $\int_0^a \frac{dx}{1+x^2} = \tan^{-1}a$; the larger a becomes, the nearer $\tan^{-1}a$ approaches to $\frac{\pi}{2}$, and by taking a sufficiently large, we can make $\tan^{-1}a$ differ as little as we please from $\frac{\pi}{2}$; hence we may write $\int_0^\infty \frac{dx}{1+x^2} = \frac{\pi}{2}$ as an abbreviation of this statement.

Similarly $\int_0^a \frac{dx}{1+x} = \log(1+a)$; and by taking a large enough we can make $\log(1+a)$ greater than any assigned quantity. Hence for abbreviation we may write

$$\int_0^\infty \frac{dx}{1+x} = \infty.$$

49. Suppose the function $\phi(x)$ to become infinite *once* between the limits a and b, namely, when $x = c$. We cannot then apply the ordinary rules of integration to $\int_a^b \phi(x)\,dx$; but we may apply those rules to

$$\int_a^{c-\mu} \phi(x)\,dx + \int_{c+\mu}^b \phi(x)\,dx$$

for any assigned value of μ however small. The limit of the last expression when μ is diminished indefinitely is called by Cauchy the *principal* value of the integral $\int_a^b \phi(x)\,dx$.

For example, let $\phi(x) = \frac{1}{c-x}$;

$$\text{then } \int_a^{c-\mu} \frac{dx}{c-x} = \log\frac{c-a}{\mu},$$

and $$\int_{c+\mu}^b \frac{dx}{c-x} = -\int_{c+\mu}^b \frac{dx}{x-c} = -\log\frac{b-c}{\mu};$$

hence the *principal* value is $\log \frac{c-a}{\mu} - \log \frac{b-c}{\mu}$, that is $\log \frac{c-a}{b-c}$.

50. The value of $\int \frac{dx}{\sqrt{(a^2-x^2)}}$ is $\sin^{-1} \frac{x}{a}$; hence

$$\int_{-a}^{a} \frac{dx}{\sqrt{(a^2-x^2)}} = \sin^{-1}(1) - \sin^{-1}(-1).$$

Students are sometimes doubtful respecting the value which is to be assigned to $\sin^{-1}(1)$ and to $\sin^{-1}(-1)$ in such a result as the above. Suppose we assume $x = a \sin \theta$; thus the integral becomes $\int d\theta$ or θ. Now x increases from $-a$ to a, hence the limits assigned to θ must be such as correspond to this range of values of x. When $x = -a$ then θ may have any value contained in the formula $(4n-1)\frac{\pi}{2}$, where n is any integer. Suppose we take the value $(4n-1)\frac{\pi}{2}$, where n is some definite integer, then corresponding to the value $x = a$ we *must* take $\theta = (4n-1)\frac{\pi}{2} + \pi$; this will be obvious on examination, because x is to change from $-a$ to $+a$, so that it *continually increases and only once passes through the value zero.*

Hence
$$\int_{-a}^{a} \frac{dx}{\sqrt{(a^2-x^2)}} = \pi.$$

As this point is frequently found to be difficult by beginners we will consider another example.

Suppose we require $\int_0^{\pi} \frac{\sec^2 \theta \, d\theta}{a^2 + \tan^2 \theta}$.

We have $\int \frac{\sec^2 \theta \, d\theta}{a^2 + \tan^2 \theta} = \frac{1}{a} \tan^{-1} \left(\frac{\tan \theta}{a} \right)$;

and as the integral is to be taken between the limits 0 and π, we must determine the values of $\tan^{-1}\left(\frac{\tan\theta}{a}\right)$ in these cases. Suppose $0, \theta_1, \theta_2, \theta_3, \ldots \theta_n, \pi$, to be a series of quantities in order of magnitude. By the nature of integration

$$\int_0^\pi u d\theta = \int_0^{\theta_1} u d\theta + \int_{\theta_1}^{\theta_2} u d\theta + \int_{\theta_2}^{\theta_3} u d\theta + \ldots + \int_{\theta_n}^{\pi} u d\theta.$$

Now each of the integrals on the right-hand side can be made as small as we please by increasing n and making two consecutive quantities as θ_r and θ_{r+1} to differ as little as we please. Hence we see that the symbol $\tan^{-1}\left(\frac{\tan\theta}{a}\right)$ must be so taken that $\tan^{-1}\left(\frac{\tan\theta_{r+1}}{a}\right) - \tan^{-1}\left(\frac{\tan\theta_r}{a}\right)$ shall diminish indefinitely when $\theta_{r+1} - \theta_r$ does so.

Therefore $\tan^{-1}\left(\frac{\tan\theta}{a}\right)$ must increase continuously with θ, and it can only pass once through an odd multiple of $\frac{\pi}{2}$ while θ passes from 0 to π. If then we take $m\pi$ for the value of $\tan^{-1}\left(\frac{\tan\theta}{a}\right)$ when $\theta = 0$, we must take $(m+1)\pi$ for the value when $\theta = \pi$; and thus the value of the integral between the assigned limits is $\frac{\pi}{a}$.

A common mistake with beginners is to take the second value the same as the first, instead of taking the second value to exceed the first by π; thus the value of the proposed integral is made to be zero, which contradicts Art. 43.

Again, suppose we require $\int_0^\pi \frac{(a - c\cos\theta)\, d\theta}{a^2 + c^2 - 2ac\cos\theta}$.

$$\int \frac{(a - c\cos\theta)\, d\theta}{a^2 + c^2 - 2ac\cos\theta} = \frac{1}{2a}\int\left\{1 + \frac{a^2 - c^2}{a^2 + c^2 - 2ac\cos\theta}\right\} d\theta.$$

Thus the required integral is $\frac{\pi}{2a} + \frac{a^2 - c^2}{2a}\int_0^\pi \frac{d\theta}{a^2 + c^2 - 2ac\cos\theta}$.

Now $\int \frac{d\theta}{a^2+c^2-2ac\cos\theta}$

$$=\int \frac{\sec^2 \frac{1}{2}\theta\, d\theta}{(a-c)^2+(a+c)^2\tan^2\frac{1}{2}\theta} = \frac{2}{a^2-c^2}\tan^{-1}\left(\frac{a+c}{a-c}\tan\tfrac{1}{2}\theta\right).$$

When taken between the assigned limits this gives $\frac{2}{a^2-c^2}\frac{\pi}{2}$ if a is greater than c, and $-\frac{2}{a^2-c^2}\frac{\pi}{2}$ if a is less than c.

Hence the value of the proposed integral is $\frac{\pi}{a}$ if a is greater than c, and zero if a is less than c.

51. Required $\int_0^{\frac{1}{2}\pi} \log\sin x\, dx$.

By equation (3) of Art. 41,

$$\int_0^{\frac{1}{2}\pi} \log\sin x\, dx = \int_0^{\frac{1}{2}\pi} \log\sin\left(\frac{\pi}{2}-x\right) dx = \int_0^{\frac{1}{2}\pi} \log\cos x\, dx.$$

Hence, putting y for the required integral,

$$2y = \int_0^{\frac{1}{2}\pi} (\log\sin x + \log\cos x)\, dx$$

$$= \int_0^{\frac{1}{2}\pi} \log(\sin x\cos x)\, dx$$

$$= \int_0^{\frac{1}{2}\pi} \log\frac{\sin 2x}{2}\, dx$$

$$= \int_0^{\frac{1}{2}\pi} \{\log\sin 2x - \log 2\}\, dx$$

$$= \int_0^{\frac{1}{2}\pi} \log\sin 2x\, dx - \frac{1}{2}\pi\log 2.$$

But putting $2x = x'$, we have

$$\int_0^{\frac{1}{2}\pi} \log\sin 2x\, dx = \tfrac{1}{2}\int_0^{\pi} \log\sin x'\, dx'$$

$$= \int_0^{\frac{1}{2}\pi} \log\sin x\, dx, \text{ by equation (4) of Art. 41};$$

therefore $$2y = y - \frac{\pi}{2} \log 2,$$

therefore $$y = \frac{\pi}{2} \log \frac{1}{2}.$$

Again, $\int_0^\pi \theta^2 \log \sin \theta \, d\theta = \int_0^\pi (\pi - \theta)^2 \log \sin \theta \, d\theta$, by equation (3) of Art. 41; therefore

$$0 = \int_0^\pi (\pi^2 - 2\pi\theta) \log \sin \theta \, d\theta,$$

therefore $$\int_0^\pi \theta \log \sin \theta \, d\theta = \frac{\pi}{2} \int_0^\pi \log \sin \theta \, d\theta = \frac{\pi^2}{2} \log \frac{1}{2}.$$

Required $\int_0^1 \frac{\log(1+x)}{1+x^2} dx$. Put $x = \tan y$, and the integral becomes $\int_0^{\frac{\pi}{4}} \log (1 + \tan y) \, dy$; but by equation (3) of Art. 41

$$\int_0^{\frac{\pi}{4}} \log (1 + \tan y) \, dy = \int_0^{\frac{\pi}{4}} \log \left\{1 + \tan \left(\frac{\pi}{2} - y\right)\right\} dy,$$

and $$1 + \tan \left(\frac{\pi}{4} - y\right) = 1 + \frac{1 - \tan y}{1 + \tan y} = \frac{2}{1 + \tan y};$$

therefore $$2 \int_0^{\frac{\pi}{4}} \log (1 + \tan y) \, dy = \frac{\pi}{4} \log 2;$$

therefore $$\int_0^1 \frac{\log(1+x)}{1+x^2} dx = \frac{\pi}{8} \log 2.$$

See *Cambridge Mathematical Journal*, Vol. III. p. 168.

52. The *remainder* after $n+1$ terms of the expansion of $\phi(a+h)$ in powers of h, may be expressed by a definite integral. For let

$$F(z) = \phi(x-z) + z\phi'(x-z) + \frac{z^2}{\lfloor 2} \phi''(x-z) \ldots\ldots + \frac{z^n}{\lfloor n} \phi^n (x-z).$$

Differentiate with respect to z, then

$$F'(z) = -\frac{z^n}{\lfloor n} \phi^{n+1}(x-z).$$

Integrate both members of this equation between the limits 0 and h; thus

$$F(h) - F(0) = -\frac{1}{\lfloor n}\int_0^h z^n \phi^{n+1}(x-z)\,dz,$$

that is,

$$\phi(x-h) + h\phi'(x-h) + \frac{h^2}{\lfloor 2}\phi''(x-h) \ldots\ldots + \frac{h^n}{\lfloor n}\phi^n(x-h) - \phi(x)$$

$$= -\frac{1}{\lfloor n}\int_0^h z^n \phi^{n+1}(x-z)\,dz.$$

Put $a+h$ for x and transpose, then

$$\phi(a+h) = \phi(a) + h\phi'(a) + \frac{h^2}{\lfloor 2}\phi''(a) \ldots\ldots + \frac{h^n}{\lfloor n}\phi^n(a)$$

$$+ \frac{1}{\lfloor n}\int_0^h z^n \phi^{n+1}(a+h-z)\,dz.$$

Thus the excess of $\phi(a+h)$ over the sum of the first $n+1$ terms of its expansion by Taylor's Theorem is expressed by the definite integral

$$\frac{1}{\lfloor n}\int_0^h z^n \phi^{n+1}(a+h-z)\,dz.$$

By means of the first result in Art. 40, we may put for this definite integral

$$\frac{\theta^n h^{n+1}}{\lfloor n}\phi^{n+1}(a+h-\theta h),$$

where θ is a proper fraction.

By means of the second result in Art. 40, we may put for this definite integral

$$\frac{1}{\lfloor n}\phi^{n+1}(a+h-\theta h)\int_0^h z^n dz,$$

or
$$\frac{h^{n+1}}{\lfloor n+1}\phi^{n+1}(a+\theta_1 h),$$

where θ_1 is also a proper fraction.

53. *Bernoulli's Series.* By integration by parts we have

$$\int \phi(x)\,dx = x\,\phi(x) - \int x\,\phi'(x)\,dx,$$

$$\int x\,\phi'(x)\,dx = \frac{x^2}{2}\,\phi'(x) - \int \frac{x^2}{2}\,\phi''(x)\,dx,$$

$$\int x^2\phi''(x)\,dx = \frac{x^3}{3}\,\phi''(x) - \int \frac{x^3}{3}\,\phi'''(x)\,dx.$$

....

Thus
$$\int \phi(x)\,dx = x\,\phi(x) - \frac{x^2}{1.2}\,\phi'(x) + \frac{x^3}{\lfloor 3}\,\phi''(x)\ldots\ldots + \frac{(-1)^{n-1}x^n}{\lfloor n}\,\phi^{n-1}(x) + \frac{(-1)^n}{\lfloor n}\int x^n\phi^n(x)\,dx.$$

Therefore,

$$\int_0^a \phi(x)\,dx = a\,\phi(a) - \frac{a^2}{1.2}\,\phi'(a) + \frac{a^3}{\lfloor 3}\,\phi''(a)\ldots\ldots + \frac{(-1)^{n-1}a^n\phi^{n-1}(a)}{\lfloor n} + \frac{(-1)^n}{\lfloor n}\int_0^a x^n\phi^n(x)\,dx.$$

This series on the right hand is called Bernoulli's series. In some cases this process might be of use in obtaining $\int_0^a \phi(x)\,dx$; for example, if $\phi(x)$ be any rational algebraical function of the $(n-1)^{\text{th}}$ degree, $\phi^n(x)$ is zero; or it might happen that $\int x^n\phi^n(x)\,dx$ could be found more easily than $\int \phi(x)\,dx$. Or again, we may require only an *approximate* value of $\int_0^a \phi(x)\,dx$ and the integral $\int_0^a x^n\phi^n(x)\,dx$ might be small enough to be neglected.

54. By adopting different methods of integrating a function, we may apparently sometimes arrive at different results. But we know (*Dif. Calc.* Art. 102) that two functions which have the same differential coefficient can only differ by a constant, so that any two results which we obtain must either be identical or differ by a constant. Take for example

$$\int (ax+b)(a'x+b')\,dx;$$

integrate by parts, thus we obtain

$$\frac{(ax+b)^2}{2a}(a'x+b') - \int \frac{a'}{2a}(ax+b)^2\,dx,$$

that is, $$\frac{(ax+b)^2(a'x+b')}{2a} - \frac{a'(ax+b)^3}{6a^2}.$$

If we integrate by parts in another way, we can obtain

$$\frac{(a'x+b')^2(ax+b)}{2a'} - \frac{a(a'x+b')^3}{6a'^2}.$$

Hence

$$\frac{(ax+b)^2\{3a(a'x+b') - a'(ax+b)\}}{6a^2}$$

and $$\frac{(a'x+b')^2\{3a'(ax+b) - a(a'x+b')\}}{6a'^2}$$

can only differ by a constant. Hence multiplying by $6a^2a'^2$ we have

$$a'^2(ax+b)^2\{3a(a'x+b') - a'(ax+b)\}$$
$$- a^2(a'x+b')^2\{3a'(ax+b) - a(a'x+b')\} = C,$$

where C is some constant. This might of course be verified by common reduction. We may easily determine the value of C; for since it is independent of x we may suppose $ax+b=0$, that is, $x = -\frac{b}{a}$; then the left-hand member becomes $(ab'-a'b)^3$, which is consequently the value of C.

Similarly from

$$\int (ax+b)\,dx + \int (a'x+b')\,dx = \int \{(a+a')\,x+b+b'\}\,dx$$

we infer

$$\frac{(ax+b)^2}{2a} + \frac{(a'x+b')^2}{2a'} = \frac{\{(a+a')\,x+b+b'\}^2}{2\,(a+a')} + \text{constant}.$$

Multiply by $2aa'(a+a')$ and then determine the constant by supposing $x=0$; thus we obtain the identity

$$a'(a+a')\,(ax+b)^2 + a\,(a+a')\,(a'x+b')^2$$
$$= aa'\,\{(a+a')\,x+b+b'\}^2 + (ba'-b'a)^2.$$

55. By $\int \phi(x)\,dx$ we indicate the function of which $\phi(x)$ is the differential coefficient; suppose this to be $\psi(x)$. Then we may require the function of which $\psi(x)$ is the differential coefficient, which we denote by $\int \psi(x)\,dx$, or by $\iint \phi(x)\,dx\,dx$, and so on. For example, the integral of e^{kx} is $\frac{1}{k}e^{kx} + C_1$, where C_1 is a constant; the integral of this is

$$\frac{1}{k^2}e^{kx} + C_1 x + C_2;$$

the integral of this is

$$\frac{1}{k^3}e^{kx} + C_1\frac{x^2}{2} + C_2 x + C_3,$$

where $\frac{C_1}{2}$ being still a constant may be denoted for simplicity by B if we please. Proceeding thus we should find as the result of integrating e^{kx} successively for n times

$$\frac{e^{kx}}{k^n} + A_1 x^{n-1} + A_2 x^{n-2} + \ldots\ldots + A_{n-1}x + A_n,$$

where $A_1, A_2, \ldots\ldots A_n$ are constants.

It is easy to express a repeated integral in terms of simple integrals. For let u be any function of x; let

$$u_1 = \int u dx; \text{ let } u_2 = \int u_1 dx; \text{ let } u_3 = \int u_2 dx;$$

and so on.

By integration by parts we have

$$u_2 = \int u_1 dx = xu_1 - \int x \frac{du_1}{dx} dx = x \int u dx - \int xu\, dx;$$

$$u_3 = \int u_2 dx = \int \left\{ x \int u dx - \int xu\, dx \right\} dx;$$

therefore by integration by parts,

$$u_3 = \frac{x^2}{2} \int u dx - \int \frac{x^2}{2} u dx - x \int xu\, dx + \int x^2 u dx$$

$$= \frac{x^2}{2} \int u dx - x \int xu\, dx + \frac{1}{2} \int x^2 u dx.$$

The general formula is

$$\underline{|n}\, u_{n+1} = x^n \int u dx - nx^{n-1} \int xu\, dx + \frac{n(n-1)}{1.2} x^{n-2} \int x^2 u dx - \ldots\ldots$$

$$\ldots\ldots + (-1)^r \frac{n(n-1)\ldots(n-r+1)}{\underline{|r}} x^{n-r} \int x^r u dx + \ldots\ldots$$

$$\ldots\ldots + (-1)^n \int x^n u dx.$$

The truth of this formula may be easily established by induction; for if we differentiate both sides we obtain a similar formula with $n - 1$ in place of n.

MISCELLANEOUS EXAMPLES.

1. $\int_0^a \frac{x^{\frac{5}{2}} dx}{\sqrt{(a-x)}} = \frac{5\pi a^3}{16}$. (Assume $x = a \sin^2 \theta$).

2. $\int_0^{2a} \frac{x dx}{\sqrt{(2ax - x^2)}} = \pi a$.

3. $\int_0^a \frac{(a^2 - e^2 x^2)\, dx}{\sqrt{(a^2 - x^2)}} = \frac{\pi a^2}{2}\left(1 - \frac{e^2}{2}\right)$.

4. $\int_0^\infty \frac{dx}{(a^2 + x^2)(b^2 + x^2)} = \frac{\pi}{2ab(a+b)}$.

5. If $\phi(x) = \phi(a + x)$, shew that

$$\int_0^{na} \phi(x)\, dx = n \int_0^a \phi(x)\, dx.$$

6. Shew that $\int_a^b \phi(x) dx = \frac{b-a}{2c} \int_{-c}^{c} \phi\left(\frac{b+a}{2} + \frac{b-a}{2c} x\right) dx$.

7. Shew that $\int_0^\pi \frac{x \sin x dx}{1 + \cos^2 x} = \frac{\pi^2}{4}$. (Change x into $\pi - x'$.)

8. Shew that $\int_0^{2a} (2ax - x^2)^{\frac{3}{2}} \text{vers}^{-1} \frac{x}{a} dx = \frac{3\pi^2 a^4}{16}$.
(Change x into $2a - x'$.)

9. Find the limit when n is infinite of

$$\frac{1}{n} + \frac{1}{\sqrt{(n^2 - 1)}} + \frac{1}{\sqrt{(n^2 - 2^2)}} + \ldots\ldots + \frac{1}{\sqrt{\{n^2 - (n-1)^2\}}}.$$

Result. $\frac{\pi}{2}$.

10. Find the limit when n is infinite of

$$\frac{\left(\frac{1}{2n}\right)^p + \left(\frac{2}{2n}\right)^p + \left(\frac{3}{2n}\right)^p + \text{... to } 2n \text{ terms}}{\left(\frac{1}{2}+\frac{1}{2n}\right)^p + \left(\frac{1}{2}+\frac{2}{2n}\right)^p + \left(\frac{1}{2}+\frac{3}{2n}\right)^p + \text{... to } n \text{ terms}}.$$

Result. $\dfrac{1}{1-\left(\frac{1}{2}\right)^{p+1}}$.

11. Find the limit when n is infinite of $\left\{\dfrac{\lfloor n}{n^n}\right\}^{\frac{1}{n}}$.

Result. $\dfrac{1}{e}$. (Take the logarithm of the expression.)

12. Shew that $\int_0^{\frac{\pi}{2}} \log \tan x \, dx = 0$.

13. Shew that $\int_0^{\frac{\pi}{2}} \sin x \log \sin x \, dx = \log 2 - 1$.

14. If $f(x)$ be positive and finite from $x = a$ to $x = a + c$, shew how to find the limit of

$$\left\{f(a) f\left(a + \frac{c}{n}\right) \text{......} f\left(a + \frac{n-1}{n} c\right)\right\}^{\frac{1}{n}}$$

when n is infinite; and prove that the limit in question is less than $\dfrac{1}{c}\int_a^{a+c} f(x)\, dx$, assuming that the geometric mean of a finite number of positive quantities which are not all equal is less than the arithmetic.

Hence prove that $e^{\int_0^1 u dx}$ is less than $\int_0^1 e^u dx$, unless u be constant from $x = 0$ to $x = 1$.

15. The value of the definite integral $\int_0^{\frac{\pi}{2}} \log(1 + n\cos^2\theta)\, d\theta$ may be found whatever positive value is given to n from the formula

$$\int_0^{\frac{\pi}{2}} \log(1 + n\cos^2\theta)\, d\theta = \frac{\pi}{4} \log\{(1+n)(1+n_1)^{\frac{1}{2}}(1+n_2)^{\frac{1}{4}}\ldots\}$$

where n, n_1, n_2,...... are quantities connected by the equation

$$n_{r+1} = \frac{n_r^2}{4(n_r + 1)}.$$

16. Shew that

$$\int e^{cx} \cos ax\, dx = \frac{e^{cx}\cos(ax - \phi)}{(a^2 + c^2)^{\frac{1}{2}}} + \text{a constant},$$

where $\tan\phi = \frac{a}{c}$. Hence shew that if $e^{cx}\cos ax$ be integrated n times successively the result is

$$\frac{e^{cx}\cos(ax - n\phi)}{(a^2 + c^2)^{\frac{n}{2}}} + C + C_1 x + C_2 x^2 \ldots\ldots + C_{n-1} x^{n-1}.$$

CHAPTER V.

DOUBLE INTEGRATION.

56. LET $\phi(x)$ denote any function of x; then we have seen that the *integral* of $\phi(x)$ is a quantity u such that $\frac{du}{dx} = \phi(x)$. The integral may also be regarded as the limit of a certain sum (see Arts. 2—6), and hence is derived the symbol $\int \phi(x)\, dx$ by which the integral is denoted. We now proceed to extend these conceptions of an integral to cases where we have more than one independent variable.

57. Suppose we have to find the value of u which satisfies the equation $\frac{d^2u}{dy\, dx} = \phi(x, y)$, where $\phi(x, y)$ is a function of the independent variables x and y. The equation may be written

$$\frac{d}{dy}\left(\frac{du}{dx}\right) = \phi(x, y),$$

or

$$\frac{dv}{dy} = \phi(x, y),$$

if $v = \frac{du}{dx}$. Thus v must be a function such that if we differentiate it with respect to y, considering x as constant, the result will be $\phi(x, y)$. We may therefore put

$$v = \int \phi(x, y)\, dy,$$

that is,

$$\frac{du}{dx} = \int \phi(x, y)\, dy.$$

Hence u must be such a function that if we differentiate it with respect to x, considering y constant, the result will be the function denoted by $\int \phi(x, y)\, dy$. Hence

$$u = \int \left\{ \int \phi(x, y)\, dy \right\} dx.$$

The method of obtaining u may be described by saying that we first integrate $\phi(x, y)$ with respect to y, and then integrate the result with respect to x.

The above expression for u may be more concisely written thus,

$$\iint \phi(x, y)\, dy\, dx, \quad \text{or} \quad \iint \phi(x, y)\, dx\, dy.$$

On this point of notation writers are not quite uniform; we shall in the present work adopt the latter form, that is, of the two symbols dx and dy we shall put dy to the right, when we consider the integration with respect to y performed before the integration with respect to x, and *vice versa*.

58. We might find u by integrating first with respect to x and then with respect to y; this process would be indicated by the equation

$$u = \iint \phi(x, y)\, dy\, dx.$$

59. Since we have thus *two methods* of finding u from the equation $\frac{d^2u}{dx\, dy} = \phi(x, y)$, it will be desirable to investigate if more than *one result* can be obtained. Suppose then that u_1 and u_2 are two functions either of which when put for u satisfies the given equation, so that

$$\frac{d^2u_1}{dx\, dy} = \phi(x, y) \quad \text{and} \quad \frac{d^2u_2}{dx\, dy} = \phi(x, y).$$

We have, by subtraction,

$$\frac{d^2u_1}{dx\, dy} - \frac{d^2u_2}{dx\, dy} = 0,$$

that is, $$\frac{d}{dx}\left(\frac{dv}{dy}\right) = 0, \text{ where } v = u_1 - u_2.$$

Now from an equation $\frac{dw}{dx}=0$ we infer that w must be a *constant*, that is, must be a *constant* so far as relates to x; in other words, w cannot be a function of x, but *may* be a function of any other variable which occurs in the question we are considering.

Thus from the equation $\frac{d}{dx}\left(\frac{dv}{dy}\right)=0$ we infer that $\frac{dv}{dy}$ cannot be a function of x, but *may* be any arbitrary function of y. Thus we may put

$$\frac{dv}{dy}=f(y).$$

By integration we deduce

$$v=\int f(y)\,dy+\text{constant}.$$

Here the constant, as we call it, must not contain y, but may contain x; we may denote it by $\chi(x)$. And $\int f(y)\,dy$ we will denote by $\psi(y)$; thus finally

$$v=\psi(y)+\chi(x).$$

Therefore two values of u which satisfy the equation $\frac{d^2u}{dx\,dy}=\phi(x,y)$ can only differ by the sum of two arbitrary functions, one of x only and the other of y only.

60. We shall now shew the connexion between double integration and summation. Let $\phi(x, y)$ be a function of x and y, which remains finite and continuous so long as x lies between the fixed values a and b, and y between the fixed values α and β. Let $a, x_1, x_2, \ldots\ldots x_{n-1}, b$ be a series of quantities in order of magnitude; also let $\alpha, y_1, y_2, \ldots\ldots y_{m-1}, \beta$ be another series of quantities in order of magnitude.

Let $\quad x_1-a=h_1,\ x_2-x_1=h_2, \ldots\ldots\ldots b-x_{n-1}=h_n;$

also let $\quad y_1-\alpha=k_1,\ y_2-y_1=k_2, \ldots\ldots\ldots \beta-y_{m-1}=k_m.$

We propose now to find the limit of the sum of a certain series in which every term is of the form

$$h_r k_s \phi(x_{r-1}, y_{s-1}),$$

where r takes all integral values between 1 and n inclusive, and s takes all integral values between 1 and m inclusive; and ultimately m and n are to be supposed infinite; also x_0 and y_0 are to be considered equivalent to a and α respectively. Thus we may take $hk\phi(x, y)$ as the type of the terms we wish to sum, or we may take $\Delta x\,\Delta y\,\phi(x, y)$ as a still more expressive symbol. The series then is

$$h_1\{k_1\phi(a, \alpha) + k_2\phi(a, y_1) + k_3\phi(a, y_2) \ldots\ldots + k_m\phi(a, y_{m-1})\}$$
$$+ h_2\{k_1\phi(x_1, \alpha) + k_2\phi(x_1, y_1) + k_3\phi(x_1, y_2) \ldots\ldots + k_m\phi(x_1, y_{m-1})\}$$
$$\ldots\ldots\ldots\ldots\ldots\ldots\ldots\ldots\ldots\ldots\ldots\ldots\ldots\ldots$$
$$+ h_n\{k_1\phi(x_{n-1}, \alpha) + k_2\phi(x_{n-1}, y_1) + \ldots\ldots\ldots + k_m\phi(x_{n-1}, y_{m-1})\}.$$

Consider one of the horizontal rows of terms which we may write

$$h_{r+1}\{k_1\phi(x_r, \alpha) + k_2\phi(x_r, y_1) + k_3\phi(x_r, y_2) \ldots\ldots + k_m\phi(x_r, y_{m-1})\}.$$

The limit of the series within the brackets when $k_1, k_2, \ldots k_m$ are indefinitely diminished is, by Art. 3,

$$\int_\alpha^\beta \phi(x_r, y)\, dy.$$

Since this is the limit of the series, we may suppose the series itself equal to

$$\int_\alpha^\beta \phi(x_r, y)\, dy + \rho_{r+1},$$

where ρ_{r+1} ultimately vanishes.

Let $\int_\alpha^\beta \phi(x_r, y)\, dy$ be denoted by $\psi(x_r)$; then add all the horizontal rows and we obtain a result which we may denote by

$$\Sigma h\,\psi(x) + \Sigma h\rho.$$

Now diminish indefinitely each term of which h is the type, then $\Sigma h\rho$ vanishes, and we have finally

$$\int_a^b \psi(x)\,dx;$$

that is,
$$\int_a^b \left\{\int_\alpha^\beta \phi(x, y)\,dy\right\} dx.$$

This is more concisely written

$$\int_a^b \int_\alpha^\beta \phi(x, y)\,dx\,dy,$$

dy being placed to the right of dx because the integration is performed first with respect to y.

61. We may again remind the student that writers are not all agreed as to the notation for double integrals. Thus we use $\int_a^b \int_\alpha^\beta \phi(x, y)\,dx\,dy$ to imply the following order of operations: integrate $\phi(x, y)$ with respect to y between the limits α and β; then integrate the result with respect to x between the limits a and b. Some writers would denote the same order of operations by $\int_a^b \int_\alpha^\beta \phi(x, y)\,dy\,dx$.

62. We might have found the limit of the sum in Art. 60 by first taking all the terms in one vertical column, and then taking all the columns. In this way we should obtain as the sum $\int_\alpha^\beta \int_a^b \phi(x, y)\,dy\,dx$; and consequently

$$\int_\alpha^\beta \int_a^b \phi(x, y)\,dy\,dx = \int_a^b \int_\alpha^\beta \phi(x, y)\,dx\,dy.$$

63. Hitherto we have integrated both with respect to x and y between constant limits; in applications of double integration, however, the limits in the first integration are often functions of the other variable. Thus, for example, the

symbol $\int_a^b \int_{\chi(x)}^{\psi(x)} \phi(x, y)\, dx\, dy$ will denote the following operations: first integrate with respect to y considering x constant; suppose $F(x, y)$ to be the integral; then by taking the integral between the assigned limits we have the result

$$F\{x, \psi(x)\} - F\{x, \chi(x)\}.$$

We have finally to obtain the integral indicated by

$$\int_a^b [F\{x, \psi(x)\} - F\{x, \chi(x)\}]\, dx.$$

The only difference which is required in the summatory process of Art. 60 is, that the quantities α, y_1, y_2,...y_{m-1} will not have the same meaning in each horizontal line. In the $(r+1)^{\text{th}}$ line, for example, that is, in

$$h_{r+1}\{k_1\phi(x_r, \alpha) + k_2\phi(x_r, y_1) + k_3\phi(x_r, y_2)\ldots\ldots + k_m\phi(x_r, y_{m-1})\},$$

we must consider α as standing for $\chi(x_r)$, and y_1, y_2,...... as a series of quantities, such that $\chi(x_r)$, y_1, y_2,......y_{m-1}, $\psi(x_r)$, are in order of magnitude, and that the difference between any consecutive two ultimately vanishes. Hence, proceeding as before, we get $\int_{\chi(x_r)}^{\psi(x_r)} \phi(x_r, y)\, dy$ for the limit of the sum of the terms in the $(r+1)^{\text{th}}$ line.

64. It is not necessary to suppose the same number of terms in all the horizontal rows; for m is ultimately made indefinitely great, so that we obtain the same expression for the limit of the $(r+1)^{\text{th}}$ line whatever may be the number of terms with which we start.

65. When the limits in the first integration are functions of the other variable we cannot perform the integrations in a different order, as in Art. 62, without special investigation to determine what the limits will then be. This question will be considered in a subsequent chapter.

66. From the definition of double integration, it follows that when the limits of both integrations are constant,

$$\iint \phi(x)\, \psi(y)\, dx\, dy = \int \phi(x)\, dx \times \int \psi(y)\, dy,$$

supposing that the limits in $\int \psi (y)\, dy$ are the same as in the integration with respect to y in the left-hand member, and the limits in $\int \phi (x)\, dx$ the same as in the integration with respect to x in the left-hand member. For the left-hand member is the limit of the sum of a series of terms, such as

$$h_r k_s \phi (x_{r-1}, y_{s-1}),$$

and the right-hand member is the limit of the product of

$$h_1 \phi (x_0) + h_2 \phi (x_1) + h_3 \phi (x_2) \ldots\ldots + h_n \phi (x_{n-1}),$$

and $$k_1 \psi (y_0) + k_2 \psi (y_1) + k_3 \psi (y_2) \ldots\ldots + k_m \psi (y_{m-1}).$$

67. The reader will now be able to extend the processes given in this chapter to *triple* integrals and to *multiple* integrals generally. The symbol

$$\int_{\xi_0}^{\xi_1} \int_{\eta_0}^{\eta_1} \int_{\zeta_0}^{\zeta_1} \phi (x, y, z)\, dx\, dy\, dz$$

will indicate that the following series of operations must be performed: integrate $\phi (x, y, z)$ with respect to z between the limits ζ_0 and ζ_1 considering x and y constant; next integrate the result with respect to y between the limits η_0 and η_1 considering x constant; lastly integrate this result with respect to x between the limits ξ_0 and ξ_1. Here ζ_0 and ζ_1 may be functions of both x and y; and η_0 and η_1 may be functions of x. This triple integral is the limit of a certain series which may be denoted by $\Sigma \phi (x, y, z)\, \Delta x\, \Delta y\, \Delta z$.

MISCELLANEOUS EXAMPLES.

Obtain the following eight integrals.

1. $\int \frac{\sqrt{x}}{\sqrt{(a^3 - x^3)}}\, dx.$ (Put $y = x^{\frac{3}{2}}$.)

Result. $\frac{2}{3} \sin^{-1} \frac{x^{\frac{3}{2}}}{a^{\frac{3}{2}}}.$

2. $\displaystyle\int \frac{x^3 dx}{(x-a)(x-b)(x-c)}$.

Result. $\displaystyle x + \frac{a^3 \log(x-a)}{(a-b)(a-c)} + \frac{b^3 \log(x-b)}{(b-a)(b-c)} + \frac{c^3 \log(x-c)}{(c-a)(c-b)}$.

3. $\displaystyle\int \frac{\tan x \, dx}{1 + m^2 \tan^2 x}$. *Result.* $\displaystyle\frac{\log(\cos^2 x + m^2 \sin^2 x)}{2(m^2-1)}$.

4. $\displaystyle\int \frac{dx}{x\sqrt{(a^{2n} + x^{2n})}}$. $\left(\text{Put } x = \frac{1}{y}\right)$.

Result. $\displaystyle\frac{1}{na^n} \log \frac{x^n}{a^n + \sqrt{(a^{2n} + x^{2n})}}$.

5. $\displaystyle\int \sec x \sec 2x \, dx$.

Result. $\displaystyle\frac{1}{\sqrt{2}} \log \frac{1 + \sqrt{2} \sin x}{1 - \sqrt{2} \sin x} - \frac{1}{2} \log \frac{1 + \sin x}{1 - \sin x}$.

6. $\displaystyle\int \frac{\tan a - \tan x}{\tan a + \tan x} dx$.

Result. $\sin 2a \log \sin(a+x) - x \cos 2a$.

7. $\displaystyle\int \frac{dx}{x^4 + a^2x^2 + a^4}$.

Result. $\displaystyle\frac{1}{4a^3} \log \frac{x^2 + ax + a^2}{x^2 - ax + a^2} + \frac{1}{2a^3\sqrt{3}} \tan^{-1} \frac{xa\sqrt{3}}{a^2 - x^2}$.

8. $\displaystyle\int \frac{(a - bx^2)\, dx}{x\sqrt{\{cx^2 - (a - bx^2)^2\}}}$. $\left(\text{Put } \frac{a}{x} + bx = y.\right)$

Result. $\displaystyle\cos^{-1} \frac{y}{\sqrt{(c + 4ab)}}$.

9. Find the limit when n is infinite of

$\displaystyle\left\{\sin\frac{\pi}{n} \sin\frac{2\pi}{n} \sin\frac{3\pi}{n} \ldots\ldots \sin\frac{n\pi - \pi}{n}\right\}^{\frac{1}{n}}$. *Result.* $\displaystyle\frac{1}{2}$.

10. Shew that

$$\int_0^1 x(\tan^{-1}x)^2\,dx = \frac{\pi}{4}\left(\frac{\pi}{4}-1\right)+\log\sqrt{2}.$$

11. Shew that

$$\int_0^a\int_0^x\int_0^{x+y} e^{x+y+z}\,dx\,dy\,dz = \frac{e^{4a}}{8}-\frac{3e^{2a}}{4}+e^a-\frac{3}{8}.$$

12. Let $A=\iint x^2\,dx\,dy,\quad B=\iint xy\,dx\,dy,\quad C=\iint y^2\,dx\,dy,$

and suppose the limits of the integrations the same in the three integrals; then prove that AC is greater than B^2.

CHAPTER VI.

LENGTHS OF CURVES.

Plane Curves. Rectangular co-ordinates.

68. Let P be any point on the curve APQ, and let x, y be its co-ordinates; let s denote the length of the arc AP measured from a fixed point A up to P;

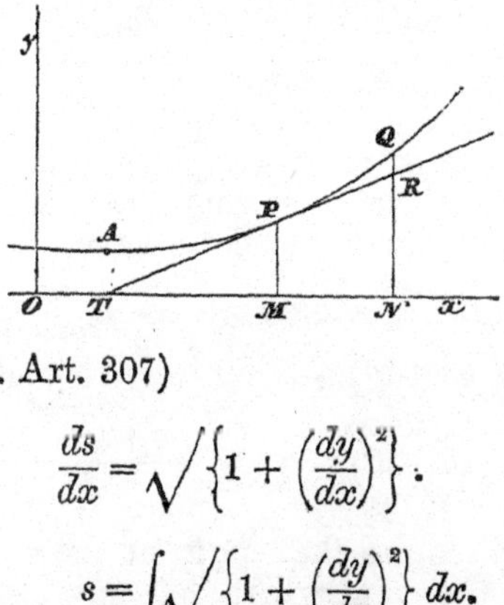

then (*Diff. Cal.* Art. 307)

$$\frac{ds}{dx} = \sqrt{\left\{1 + \left(\frac{dy}{dx}\right)^2\right\}}.$$

Hence $$s = \int \sqrt{\left\{1 + \left(\frac{dy}{dx}\right)^2\right\}}\, dx.$$

From the equation to the curve we may express $\frac{dy}{dx}$ in terms of x, and thus by integration s becomes known.

69. The process of finding the length of a curve is called the *rectification of the curve*, because we may suppose the question to be this: find a *right line* equal in length to any assigned portion of the curve.

In the preceding article we have shewn that the length of an arc of a curve will be known if a certain integral can be obtained. It may happen in many cases that this integral cannot be obtained. Whenever the length of an arc of a curve can be expressed in terms of one or both of the co-ordinates of the variable extremity of the arc, the curve is said to be *rectifiable*.

70. *Application to the Parabola.*

The equation to the parabola is $y = \sqrt{(4ax)}$; hence

$$\frac{dy}{dx} = \sqrt{\frac{a}{x}}, \quad \frac{ds}{dx} = \sqrt{\left(\frac{x+a}{x}\right)};$$

thus $$s = \int \sqrt{\left(\frac{x+a}{x}\right)} dx \qquad \text{(See Ex. 6, p. 19.)}$$

$$= \sqrt{(ax + x^2)} + a \log \{\sqrt{x} + \sqrt{(a+x)}\} + C.$$

Here C denotes some *constant* quantity, that is, some quantity which does not depend upon x; its value will depend upon the position of the fixed point from which the arc s is measured. If we measure from the vertex then s vanishes with x; hence to determine C we have

$$a \log \sqrt{a} + C = 0;$$

and thus $$s = \sqrt{(ax + x^2)} + a \log \{\sqrt{x} + \sqrt{(a+x)}\} - a \log \sqrt{a}$$

$$= \sqrt{(ax + x^2)} + a \log \frac{\sqrt{x} + \sqrt{(a+x)}}{\sqrt{a}}.$$

If then we require the length of the curve measured from the vertex to the point which has any assigned abscissa, we have only to put that assigned abscissa for x in the last expression. Thus, for example, for an extremity of the latus rectum $x = a$; hence the length of the arc between the vertex and one extremity of the latus rectum is

$$a\sqrt{2} + a \log (1 + \sqrt{2}).$$

71. In the preceding article we have found the value of the constant C, but in applying the formula to ascertain the lengths of assigned portions of curves this is not necessary.

For suppose it required to find the length of the arc of a curve measured from the point whose abscissa is x_1 up to the point whose abscissa is x_2. Let $\psi(x)$ denote the integral of $\sqrt{\left\{1+\left(\frac{dy}{dx}\right)^2\right\}}$, and let s_1 and s_2 be the lengths of arcs of the curve measured from any fixed point up to the points whose abscissæ are x_1 and x_2 respectively, so that $s_2 - s_1$ is the required length; then

$$s = \int \sqrt{\left\{1+\left(\frac{dy}{dx}\right)^2\right\}}\, dx = \psi(x) + C;$$

hence $$s_1 = \psi(x_1) + C; \quad s_2 = \psi(x_2) + C;$$

therefore $$s_2 - s_1 = \psi(x_2) - \psi(x_1).$$

Hence to find the required length we have to put x_1 and x_2 successively for x in $\psi(x)$ and subtract the first result from the second. Thus we need not take any notice of the constant C; in fact our result may be written

$$s_2 - s_1 = \int_{x_1}^{x_2} \sqrt{\left\{1+\left(\frac{dy}{dx}\right)^2\right\}}\, dx.$$

72. *Application to the Cycloid.*

In the cycloid, if the origin be at the vertex and the axis of y the tangent at that point, we have (*Diff. Cal.* Art. 358)

$$\frac{ds}{dx} = \sqrt{\left(\frac{2a}{x}\right)};$$

therefore $$s = \sqrt{(8ax)} + C.$$

The constant will be zero if we measure the arc s from the vertex.

Conversely if $s = \sqrt{(8ax)} + C$ we infer that the curve is a cycloid. And more generally if we have

$$s + A = \sqrt{(B + C_1 x + C_2 y)},$$

where A, B, C_1, and C_2 are constants, we infer that the curve is a cycloid. For by suitable changes in the origin and axes the last equation can be put in the form

$$s = \sqrt{(8ax)} + C.$$

73. *Application to the Catenary.*

The equation to the catenary is $y = \frac{c}{2}\left(e^{\frac{x}{c}} + e^{-\frac{x}{c}}\right)$; hence

$$\frac{dy}{dx} = \tfrac{1}{2}\left(e^{\frac{x}{c}} - e^{-\frac{x}{c}}\right), \quad \frac{ds}{dx} = \tfrac{1}{2}\left(e^{\frac{x}{c}} + e^{-\frac{x}{c}}\right);$$

thus $$s = \tfrac{1}{2}\int\left(e^{\frac{x}{c}} + e^{-\frac{x}{c}}\right) dx = \frac{c}{2}\left(e^{\frac{x}{c}} - e^{-\frac{x}{c}}\right) + C.$$

The constant will be zero if we measure the arc s from the point for which $x = 0$.

74. *Application to the Curve given by the equation*

$$x^{\frac{2}{3}} + y^{\frac{2}{3}} = a^{\frac{2}{3}}.$$

Here $$\frac{dy}{dx} = -\frac{y^{\frac{1}{3}}}{x^{\frac{1}{3}}}, \quad \frac{ds}{dx} = \left(\frac{x^{\frac{2}{3}} + y^{\frac{2}{3}}}{x^{\frac{2}{3}}}\right)^{\frac{1}{2}} = \frac{a^{\frac{1}{3}}}{x^{\frac{1}{3}}};$$

thus $$s = a^{\frac{1}{3}}\int\frac{dx}{x^{\frac{1}{3}}} = \frac{3a^{\frac{1}{3}}x^{\frac{2}{3}}}{2} + C.$$

The constant will be zero if we measure the arc from the point for which $x = 0$. The curve is an hypocycloid in which the radius of the revolving circle is one-fourth of the radius of the fixed circle. $\left(\text{See } \textit{Diff. Cal.} \text{ Art. 360, and put } b = \frac{a}{4}\right)$.

75. In the same way as the result in Art. 68 is obtained we may shew that

$$s = \int\sqrt{\left\{1 + \left(\frac{dx}{dy}\right)^2\right\}}\, dy.$$

Or we may derive this result from the former thus;

$$\int\sqrt{\left\{1+\left(\frac{dy}{dx}\right)^2\right\}}\,dx=\int\sqrt{\left\{1+\left(\frac{dy}{dx}\right)^2\right\}}\,\frac{dx}{dy}\,dy$$

$$=\int\sqrt{\left\{1+\left(\frac{dx}{dy}\right)^2\right\}}\,dy.$$

From the equation to the curve we may express $\frac{dx}{dy}$ in terms of y, and thus by integration s becomes known. In some cases this formula may be more convenient than that in Art. 68.

76. *Application to the Logarithmic Curve.*

The equation to this curve is $y=ba^x$, or $y=be^{\frac{x}{c}}$ if we suppose $a=e^{\frac{1}{c}}$; thus $x=c\log\frac{y}{b}$,

therefore $$\frac{dx}{dy}=\frac{c}{y},\qquad \frac{ds}{dy}=\frac{\sqrt{(c^2+y^2)}}{y},$$

and $$s=\int\frac{\sqrt{(c^2+y^2)}}{y}\,dy=\int\frac{c^2dy}{y\sqrt{(c^2+y^2)}}+\int\frac{y\,dy}{\sqrt{(c^2+y^2)}}.$$

The latter integral is $\sqrt{(c^2+y^2)}$; the former is

$$c\log\frac{y}{c+\sqrt{(c^2+y^2)}},\quad \text{(Art. 14).}$$

Hence $$s=c\log\frac{y}{c+\sqrt{(c^2+y^2)}}+\sqrt{(c^2+y^2)}+C.$$

77. If x and y are each functions of a third variable t, we have (*Dif. Cal.* Art. 307)

$$\frac{ds}{dt}=\sqrt{\left\{\left(\frac{dx}{dt}\right)^2+\left(\frac{dy}{dt}\right)^2\right\}};$$

thus $$s=\int\sqrt{\left\{\left(\frac{dx}{dt}\right)^2+\left(\frac{dy}{dt}\right)^2\right\}}\,dt.$$

78. The equation to the ellipse is $\frac{x^2}{a^2}+\frac{y^2}{b^2}=1$. We may

therefore assume $x = a \sin\phi$, $y = b \cos\phi$, so that ϕ is the complement of the *excentric angle* (*Plane Co-ordinate Geometry*, Art. 168). Therefore, by the preceding article,

$$\frac{ds}{d\phi} = \sqrt{(a^2 \cos^2\phi + b^2 \sin^2\phi)},$$

and $$s = \int \sqrt{(a^2 \cos^2\phi + b^2 \sin^2\phi)}\, d\phi = a \int \sqrt{(1 - e^2 \sin^2\phi)}\, d\phi.$$

The exact integral cannot be obtained; we may however expand $\sqrt{(1 - e^2 \sin^2\phi)}$ in a series, so that

$$s = a \int \{1 - \tfrac{1}{2} e^2 \sin^2\phi - \frac{1 \,.\, 1}{2 \,.\, 4} e^4 \sin^4\phi - \frac{1 \,.\, 1 \,.\, 3}{2 \,.\, 4 \,.\, 6} e^6 \sin^6\phi \ldots\ldots\} d\phi$$

and each term can be integrated separately. To obtain the length of the elliptic quadrant we must integrate between the limits 0 and $\frac{\pi}{2}$.

Plane Curves. Polar Co-ordinates.

79. Let r, θ be the polar co-ordinates of any point of a curve, and s the length of the arc measured from any fixed point up to this point; then (*Dif. Cal.* Art. 311)

$$\frac{ds}{d\theta} = \sqrt{\left\{r^2 + \left(\frac{dr}{d\theta}\right)^2\right\}};$$

hence $$s = \int \sqrt{\left\{r^2 + \left(\frac{dr}{d\theta}\right)^2\right\}}\, d\theta.$$

80. *Application to the Spiral of Archimedes.*

In this curve $r = a\theta$, thus $\dfrac{dr}{d\theta} = a$;

hence $$s = \int \sqrt{(r^2 + a^2)}\, d\theta = a \int \sqrt{(1 + \theta^2)}\, d\theta$$

$$= \frac{a\theta}{2} \sqrt{(1 + \theta^2)} + \frac{a}{2} \log \{\theta + \sqrt{(1 + \theta^2)}\} + C.$$

The constant will be zero if we measure the arc s from the pole, that is, from the point where $\theta = 0$.

81. *Application to the Cardioide.*

The equation to this curve is $r = a(1 + \cos\theta)$; thus

$$s = \int \sqrt{\{a^2(1 + \cos\theta)^2 + a^2 \sin^2\theta\}}\, d\theta = a\int \sqrt{(2 + 2\cos\theta)}\, d\theta$$

$$= 2a\int \cos\frac{\theta}{2}\, d\theta = 4a \sin\frac{\theta}{2} + C.$$

The constant will be zero if we measure the arc s from the point for which $\theta = 0$, that is, from the point where the curve crosses the initial line.

The length of that part of the curve which is comprised between the initial line and a line through the pole at right angles to the initial line is $4a \sin\frac{\pi}{4}$. The length of half the perimeter of the curve is $4a \sin\frac{\pi}{2}$, that is, $4a$.

82. Suppose we require the length of the complete perimeter of the cardioide; we might at first suppose that it would be equal to $2a\int_0^{2\pi} \cos\frac{\theta}{2}\, d\theta$; but this would give zero as the result, which is obviously inadmissible. The reason of this may be easily seen; we have in fact shewn that

$$\frac{ds}{d\theta} = a\sqrt{(2 + 2\cos\theta)},$$

and this ought not to be put equal to $2a \cos\frac{\theta}{2}$ but to $\pm 2a \cos\frac{\theta}{2}$, and the proper sign should be determined in any application of the formula. Now by s we understand a positive quantity, and we may measure s so that it increases with θ, and thus $\frac{ds}{d\theta}$ is positive. Hence when $\cos\frac{\theta}{2}$ is positive, we take the upper sign and put $\frac{ds}{d\theta} = 2a \cos\frac{\theta}{2}$; when $\cos\frac{\theta}{2}$ is negative, we take the lower sign and put $\frac{ds}{d\theta} = -2a \cos\frac{\theta}{2}$. Hence the length of the complete perimeter is not $2a\int_0^{2\pi}\cos\frac{\theta}{2}\,d\theta$, but $2a\int_0^{\pi}\cos\frac{\theta}{2}\,d\theta - 2a\int_{\pi}^{2\pi}\cos\frac{\theta}{2}\,d\theta$, that is, $8a$. This result might have been anticipated, for it will be obvious from the symmetry of the figure that the length of the complete perimeter is double the length of the part which is situated on one side of the initial line, and this was shewn to be $4a$ in the preceding article.

83. It may sometimes be more convenient to find the length of a curve from the formula

$$s = \int\sqrt{\left\{r^2\left(\frac{d\theta}{dr}\right)^2 + 1\right\}}\,dr,$$

which follows immediately from that in Art. 79.

84. *Application to the Logarithmic Spiral.*

The equation to this curve is $r = ba^{\theta}$, or $r = be^{\frac{\theta}{c}}$ if we suppose $a = e^{\frac{1}{c}}$; thus $\theta = c\log\frac{r}{b}$; therefore $\frac{d\theta}{dr} = \frac{c}{r}$ and

$$s = \int\sqrt{(1+c^2)}\,dr = \sqrt{(1+c^2)}\,r + C.$$

Thus the length of the portion of the curve which has r_1 and r_2 for the radii vectores of its extreme points is

$$\int_{r_1}^{r_2}\sqrt{(1+c^2)}\,dr, \text{ that is, } \sqrt{(1+c^2)}\,(r_2 - r_1).$$

The angle between the radius vector and the corresponding tangent at any point of this curve is constant (*Dif. Cal.* Art. 354); and if that angle be denoted by α we have $c = \tan\alpha$; thus $\sqrt{(1+c^2)} = \sec\alpha$; therefore $\frac{ds}{dr} = \sec\alpha$, and $s = r\sec\alpha + C$. Hence $(r_2 - r_1)\sec\alpha$ is the length of the portion mentioned above.

Formulæ involving the radius vector and perpendicular.

85. Let ϕ be the angle between the radius vector r of any point of a curve and the tangent at that point; then $\cos\phi = \dfrac{dr}{ds}$ (*Dif. Cal.* Art. 310). Let p be the perpendicular from the pole on the same tangent; then

$$\sin\phi = \frac{p}{r}, \text{ therefore } \cos\phi = \frac{\sqrt{(r^2 - p^2)}}{r};$$

thus
$$\frac{dr}{ds} = \frac{\sqrt{(r^2 - p^2)}}{r};$$

therefore
$$\frac{ds}{dr} = \frac{r}{\sqrt{(r^2 - p^2)}}, \qquad \text{and} \qquad s = \int \frac{r\,dr}{\sqrt{(r^2 - p^2)}}.$$

86. *Application to the Epicycloid.*

With the notation and figure in *Dif. Cal.* Art. 360, it may be shewn that the equation to the tangent to the epicycloid at P is

$$y' - y = -\frac{\cos\theta - \cos\dfrac{a+b}{b}\theta}{\sin\theta - \sin\dfrac{a+b}{b}\theta}(x' - x),$$

where x and y are the co-ordinates of P, and x' and y' the variable co-ordinates. Hence it will be found that the perpendicular p from the origin on the tangent at P is given by

$$p = (a + 2b)\sin\frac{a\theta}{2b};$$

also
$$r^2 = a^2 + 4b(a + b)\sin^2\frac{a\theta}{2b};$$

thus
$$p^2 = \frac{c^2(r^2 - a^2)}{c^2 - a^2}, \text{ where } c = a + 2b.$$

Hence, by Art. 85,

$$s = \frac{\sqrt{(c^2 - a^2)}}{a}\int \frac{r\,dr}{\sqrt{(c^2 - r^2)}} = -\frac{\sqrt{(c^2 - a^2)}}{a}\sqrt{(c^2 - r^2)} + C.$$

At a cusp $r=a$, and at a vertex $r=c$; thus the length of the portion of the curve between a cusp and the adjacent vertex is

$$\frac{\sqrt{(c^2-a^2)}}{a}\int_a^c \frac{rdr}{\sqrt{(c^2-r^2)}}, \text{ that is } \frac{c^2-a^2}{a}, \text{ that is } \frac{4b(a+b)}{a}.$$

Hence the length of the portion between two consecutive cusps is $\dfrac{8b(a+b)}{a}$.

87. A remark may be made here similar to that in Art. 82. If we apply the formula

$$s=-\frac{\sqrt{(c^2-a^2)}}{a}\sqrt{(c^2-r^2)}+C$$

to find the length between two consecutive cusps, we arrive at the result zero, since $r=a$ at both limits. The reason is that we have used the formula

$$\frac{ds}{dr}=\frac{\sqrt{(c^2-a^2)}}{a}\frac{r}{\sqrt{(c^2-r^2)}}$$

while the true formula is

$$\frac{ds}{dr}=\pm\frac{\sqrt{(c^2-a^2)}}{a}\frac{r}{\sqrt{(c^2-r^2)}}.$$

Since s may be taken to increase continually, it follows that $\frac{ds}{dr}$ is positive when r is increasing, and negative when r is diminishing. Now in passing along the curve from a cusp to the adjacent vertex r increases, thus $\frac{ds}{dr}$ is positive, and we should take the *upper* sign in the formula for $\frac{ds}{dr}$; then in passing from the vertex to the next cusp r diminishes, thus $\frac{ds}{dr}$ is negative, and the *lower* sign must be taken. Hence the length from one cusp to the next cusp is

$$=\frac{\sqrt{(c^2-a^2)}}{a}\int_a^c \frac{rdr}{\sqrt{(c^2-r^2)}}-\frac{\sqrt{(c^2-a^2)}}{a}\int_c^a \frac{rdr}{\sqrt{(c^2-r^2)}}$$

$$=\frac{2\sqrt{(c^2-a^2)}}{a}\int_a^c \frac{rdr}{\sqrt{(c^2-r^2)}}=\frac{8b\,(a+b)}{a}.$$

88. From what is stated in the preceding article, it appears that if the arc s begin at a vertex the proper formula is

$$\frac{ds}{dr}=-\frac{\sqrt{(c^2-a^2)}}{a}\frac{r}{\sqrt{(c^2-r^2)}},$$

therefore $s=-\dfrac{\sqrt{(c^2-a^2)}}{a}\displaystyle\int\frac{rdr}{\sqrt{(c^2-r^2)}}=\frac{\sqrt{(c^2-a^2)}}{a}\sqrt{(c^2-r^2)}.$

No constant is required since we begin to measure at the point for which $r=c$; the formula holds for values of s less than $\dfrac{4b\,(a+b)}{a}$.

It may be observed that thus

$$s=\frac{c^2-a^2}{a^2}\sqrt{(r^2-p^2)}.$$

89. Similarly for the hypocycloid we may shew that

$$p^2=\frac{c^2\,(a^2-r^2)}{a^2-c^2}\text{, where } c=a-2b.$$

Suppose c^2 less than a^2; then we may shew that

$$\frac{ds}{dr}=\pm\frac{\sqrt{(a^2-c^2)}}{a}\frac{r}{\sqrt{(r^2-c^2)}},$$

and thus s may be found. The length of the curve between two adjacent cusps is $\dfrac{8b\,(a-b)}{a}$.

Next suppose c^2 greater than a^2; then we should write the value of $\dfrac{ds}{dr}$ thus,

$$\frac{ds}{dr}=\pm\frac{\sqrt{(c^2-a^2)}}{a}\frac{r}{\sqrt{(c^2-r^2)}};$$

in this case b is greater than a, and we shall find the length of the curve between two adjacent cusps to be $\frac{8b(b-a)}{a}$.

When $a = 2b$ we have $c = 0$ and $p = 0$; in this case the hypocycloid becomes a straight line coinciding with a diameter of the fixed circle.

If $a = b$ we have $c^2 = a^2$; in this case the denominator in the value of p^2 vanishes; it will be found that the hypocycloid is then reduced to a point, and $r = a$.

It may be shewn as in Art. 88, that if s be measured from a vertex to a point not beyond the adjacent cusp, we have

$$s = \pm \frac{c^2 - a^2}{a^2} \sqrt{(r^2 - p^2)},$$

the upper or lower sign being taken according as c is greater or less than a.

Formulæ involving the Perpendicular and its Inclination.

90. Another method of expressing the length of a curve is worthy of notice.

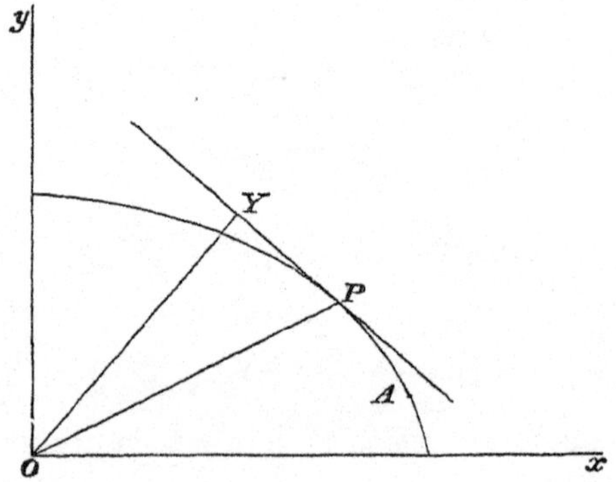

Let P be a point in a curve; x, y its co-ordinates. Let s be the length of the arc measured from a fixed point A up to P. Draw OY a perpendicular from the origin O on the tangent at P, suppose $OY = p$, $PY = u$, $YOx = \theta$; then

$$p = x \cos\theta + y \sin\theta,$$
$$u = x \sin\theta - y \cos\theta,$$
$$\frac{dy}{dx} = -\cot\theta, \quad \frac{ds}{dx} = -\operatorname{cosec}\theta;$$

therefore

$$\frac{dp}{d\theta} = -x\sin\theta + y\cos\theta + \cos\theta\frac{dx}{d\theta} + \sin\theta\frac{dy}{d\theta} = -u,$$
$$\frac{d^2p}{d\theta^2} = -\frac{du}{d\theta} = -x\cos\theta - y\sin\theta - \sin\theta\frac{dx}{d\theta} + \cos\theta\frac{dy}{d\theta}$$
$$= -p - \operatorname{cosec}\theta\frac{dx}{d\theta} = -p + \frac{ds}{d\theta};$$

therefore, by integration,

$$\frac{dp}{d\theta} = -\int p\,d\theta + s,$$

therefore $$s = \frac{dp}{d\theta} + \int p\,d\theta;$$

this may also be written

$$s + u = \int p\,d\theta.$$

Suppose s_1 and u_1 the values of s and u when θ has the value θ_1, and s_2 and u_2 their values when θ has the value θ_2, then

$$s_2 - s_1 + u_2 - u_1 = \int_{\theta_1}^{\theta_2} p\,d\theta.$$

We have measured u in the direction of revolution from P and have taken it as positive in this case; when u is negative it will indicate that Y is on the other side of P.

The preceding results may be used for different purposes, among which two may be noticed.

(1) To determine the length of any portion of a curve when the equation to the curve is given; for from that equation together with $\frac{dy}{dx} = -\cot\theta$ we can find x and y in terms of θ, and therefore p which is equal to $x\cos\theta + y\sin\theta$; then s may be found from the equation

$$s = \frac{dp}{d\theta} + \int p\,d\theta.$$

(2) To find a curve such that by means of its arc a proposed integral may be represented; for if the proposed integral be $\int p d\theta$, where p is a function of θ, the required curve is found by eliminating θ between the equations

$$x = p\cos\theta - \frac{dp}{d\theta}\sin\theta, \quad y = p\sin\theta + \frac{dp}{d\theta}\cos\theta,$$

and then the integral may be represented by $s - \dfrac{dp}{d\theta}$.

This article has been derived from Hymers's *Integral Calculus*, Art. 136.

91. The results of the preceding article may be obtained in another way. Let ρ denote the radius of curvature of the

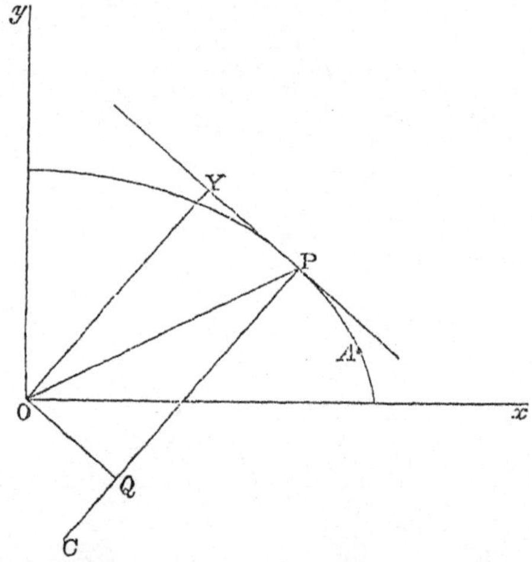

curve at P; let $OP = r$, and let s, u, and θ have the same meaning as before, then from the Differential Calculus we have

$$\rho = \frac{ds}{d\theta}, \quad \text{and } \rho = r\frac{dr}{dp}, \quad \text{therefore } \frac{dp}{d\theta} = r\frac{dr}{ds}.$$

Also $$PY = r\cos OPY = -r\frac{dr}{ds};$$

therefore $$\frac{dp}{d\theta} = -PY = -u.$$

Let PC be the radius of curvature at P; draw OQ perpendicular to PC. The locus of C is the evolute of the curve AP; and QC is with respect to this locus what PY is with respect to the locus of P. Let θ', p' be the polar co-ordinates of Q, and let $QC = u'$; then

$$\theta' = \theta - \frac{\pi}{2} \text{ and } p' = u.$$

And $$QC = u' = -\frac{dp'}{d\theta'} = -\frac{dp'}{d\theta} = -\frac{du}{d\theta} = \frac{d^2p}{d\theta^2}.$$

Also $$\rho = PQ + QC = p + u' = p + \frac{d^2p}{d\theta^2};$$

but $$\rho = \frac{ds}{d\theta}, \text{ therefore } s = \frac{dp}{d\theta} + \int p\,d\theta.$$

From the value of PY we can obtain an easy proof of a theorem of some interest in the Differential Calculus (*Dif. Cal.* Art. 329). Let p_1 denote the perpendicular from O on the locus of Y; then (*Dif. Cal.* Art. 284)

$$\frac{1}{p_1^2} = \frac{1}{p^2} + \frac{1}{p^4}\left(\frac{dp}{d\theta}\right)^2,$$

since p is the radius vector of Y. Thus

$$\frac{1}{p_1^2} = \frac{1}{p^2} + \frac{u^2}{p^4} = \frac{p^2 + u^2}{p^4} = \frac{r^2}{p^4};$$

therefore $$p_1 = \frac{p^2}{r}.$$

A particular case of the formula

$$s_2 - s_1 + u_2 - u_1 = \int_{\theta_1}^{\theta_2} p\,d\theta$$

should be noticed. Suppose we take a *complete* oval curve without singular points; then $\theta_2 = \theta_1 + 2\pi$, and $u_2 = u_1$; thus the complete perimeter of the curve is $\int_{\theta_1}^{\theta_1 + 2\pi} p\,d\theta$.

92. *Application to the Ellipse.*

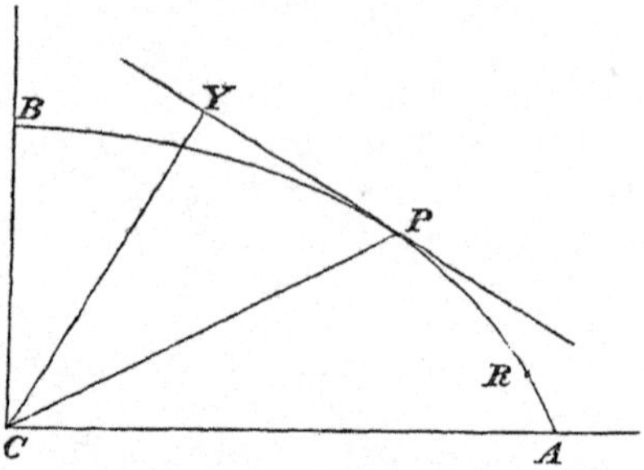

Let APB be a quadrant of an ellipse, CY the perpendicular on the tangent at P; let $ACY=\theta$. Then (*Plane Coordinate Geometry*, Art. 196) $CY=a\sqrt{(1-e^2\sin^2\theta)}$;

therefore $$AP+PY=a\int\sqrt{(1-e^2\sin^2\theta)}\,d\theta,$$

the constant to be added to the integral is supposed to be so taken that the integral may vanish with θ. If R be a point such that its excentric angle is $\frac{\pi}{2}-\theta$, we have, by Art. 78,

$$BR=a\int\sqrt{(1-e^2\sin^2\theta)}\,d\theta;$$

thus $$AP+PY=BR\ldots\ldots\ldots\ldots\ldots\ldots\ldots(1).$$

And $$PY=-\frac{dp}{d\theta}=\frac{ae^2\sin\theta\cos\theta}{\sqrt{(1-e^2\sin^2\theta)}}.$$

Let x be the abscissa of P; then by Art. 90,

$$x=p\cos\theta-\frac{dp}{d\theta}\sin\theta$$

$$=a\sqrt{(1-e^2\sin^2\theta)}\cos\theta+\frac{ae^2\sin^2\theta\cos\theta}{\sqrt{(1-e^2\sin^2\theta)}}=\frac{a\cos\theta}{\sqrt{(1-e^2\sin^2\theta)}}.$$

Thus $PY=e^2x\sin\theta$; and if x' be the abscissa of R we have $x'=a\cos\left(\frac{\pi}{2}-\theta\right)$ so that $PY=\frac{e^2xx'}{a}$. Thus (1) may be written

$$BR-AP=\frac{e^2}{a}xx'\ldots\ldots\ldots\ldots\ldots\ldots\ldots(2);$$

this result is called Fagnani's Theorem.

From the ascertained values of x and x' we have

$$x^2 = \frac{a^2 - a^2 \sin^2 \theta}{1 - e^2 \sin^2 \theta} = \frac{a^2 - x'^2}{1 - \frac{e^2 x'^2}{a^2}};$$

therefore $$e^2x^2x'^2 - a^2 (x^2 + x'^2) + a^4 = 0.$$

Thus the equation which connects x and x' involves these quantities *symmetrically;* hence from (2) we can infer that

$$BP - AR = \frac{e^2}{a} xx'.$$

This is also obvious from the figure.

We may observe that the value of PY may be obtained more simply by means of a known property of the ellipse. For suppose the normal at P to be drawn meeting CA in G; and through P draw a line parallel to CA meeting CY in Q. Then $PQ = CG = e^2x$, by the nature of the ellipse; and

$$PY = PQ \sin \theta = e^2 x \sin \theta.$$

93. *Application to the Hyperbola.*

Let C be the centre and A the vertex of an hyperbola, CY the perpendicular on the tangent at P. Let $ACY = \theta$ and $CY = p$; then it may be proved that

$$PY - AP = a \int \sqrt{(1 - e^2 \sin^2 \theta)}\, d\theta.$$

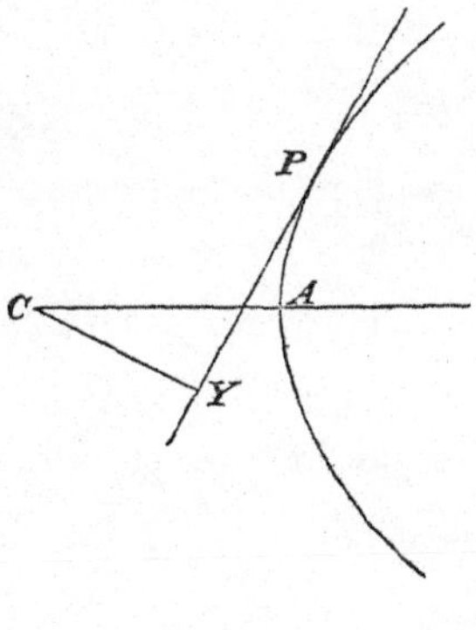

This may be proved in the same manner as the corresponding result of the preceding article; we may either make the requisite changes of sign in the formulæ of Art. 90, which are produced by difference of figure; or we may begin from the beginning again in the manner of that article. The constant to be added to the integral is supposed to be so taken that the integral may vanish with θ.

Suppose α the greatest value which θ can have, then (*Plane Co-ordinate Geometry*, Art. 257) $\cot \alpha = \sqrt{(e^2 - 1)}$. When P moves off to an infinite distance $PY - AP$ becomes the difference between the length of the asymptote from C and the infinite hyperbolic arc from A. Thus this difference is

$$a \int_0^\alpha \sqrt{(1 - e^2 \sin^2 \theta)}\, d\theta.$$

Inverse questions on the lengths of Curves.

94. In the preceding articles we have shewn how the length of an arc of a known curve is to be found in terms of the abscissa of its variable extremity; we will now briefly notice the inverse problem, to find a curve such that the arc shall be a given function of the abscissa of its variable extremity.

Suppose $\phi(x)$ the given function; then $s = \phi(x)$;

therefore $$\phi'(x) = \frac{ds}{dx} = \sqrt{\left\{1 + \left(\frac{dy}{dx}\right)^2\right\}};$$

thus $$\frac{dy}{dx} = [\{\phi'(x)\}^2 - 1]^{\frac{1}{2}},$$

and $$y = \int [\{\phi'(x)\}^2 - 1]^{\frac{1}{2}}\, dx.$$

95. As an example of the preceding method, suppose $\phi(x) = \sqrt{(4cx)}$; thus $\phi'(x) = \sqrt{\frac{c}{x}}$; therefore

$$y = \int \left[\frac{c}{x} - 1\right]^{\frac{1}{2}} dx = \int \frac{(c-x)\,dx}{\sqrt{(cx-x^2)}}$$

$$= \int \frac{\left(\frac{c}{2} - x\right) dx}{\sqrt{(cx-x^2)}} + \frac{c}{2}\int \frac{dx}{\sqrt{(cx-x^2)}}$$

$$= \sqrt{(cx-x^2)} + \frac{c}{2}\,\text{vers}^{-1}\frac{2x}{c} + C.$$

We may write y' for $y - C$ and thus we find that the curve is a cycloid. (*Dif. Cal.* Art. 358.)

96. For another example suppose $\phi(x) = a \log x$; thus $\phi'(x) = \frac{a}{x}$.

Here $$y = \int \sqrt{\left(\frac{a^2}{x^2} - 1\right)}\,dx = \int \frac{(a^2 - x^2)\,dx}{x\sqrt{(a^2-x^2)}}$$

$$= \int \frac{a^2\,dx}{x\sqrt{(a^2-x^2)}} - \int \frac{x\,dx}{\sqrt{(a^2-x^2)}}$$

$$= a \log \frac{x}{a + \sqrt{(a^2-x^2)}} + \sqrt{(a^2-x^2)} + C.$$

Involutes and Evolutes.

97. We may express the length of an arc of a curve without integration when we know the equation to the involute of the curve. Suppose s' to represent the length of an arc of a curve, ρ the radius of curvature at that point of the involute which corresponds to the variable extremity of s', then (*Dif. Cal.* Art. 331) $s' \pm \rho = l$, where l is a constant. If the equation to the involute is known, ρ can be found in terms of the co-ordinates of the point in the involute; then these co-ordinates can be expressed in terms of the co-ordinates of the corresponding point of the evolute, and thus s' is known. By this method we have to perform the processes of differentiation and algebraical reduction instead of integration.

98. *Application to the Evolute of the Parabola.*

Take for the involute the parabola which has for its equation $y^2 = 4ax$; let x', y' be the co-ordinates of the point of the evolute which corresponds to the point (x, y) on the parabola. Then by the ordinary methods (*Dif. Cal.* Art. 330) we have

$$x' = 2a + 3x, \quad y' = -\frac{y^3}{4a^2},$$

and $$\rho = 2a\left(\frac{a+x}{a}\right)^{\frac{3}{2}}.$$

Thus we shall obtain for the equation to the evolute

$$27ay'^2 = 4(x' - 2a)^3;$$

and $$\rho = 2a\left(\frac{x'+a}{3a}\right)^{\frac{3}{2}};$$

therefore $$s' \pm 2a\left(\frac{x'+a}{3a}\right)^{\frac{3}{2}} = l.$$

Suppose we measure s' from the point for which $x' = 2a$, that is, from the point which corresponds to the vertex of the parabola; then we see that s' increases with x', so that we must take the lower sign in the last equation; also by supposing $x' = 2a$ and $s' = 0$ we find $l = -2a$; thus

$$s' = 2a\left(\frac{x'+a}{3a}\right)^{\frac{3}{2}} - 2a.$$

This value of s' may also be obtained by the application of the ordinary method of integration.

99. When the length of the arc of a curve is known in terms of the co-ordinates of its variable extremity, the equation to the involute can be found by ordinary processes of elimination.

For we have (*Dif. Cal.* Art. 331)

$$\frac{\frac{dx'}{dx}}{x' - x} = \pm\frac{1}{\rho}\frac{ds'}{dx},$$

where the accented letters refer to a point in a curve, and the unaccented letters to the corresponding point in the involute. Thus

$$x = x' \mp \rho \frac{dx'}{ds'} \text{.......................} (1).$$

Similarly $$y = y' \mp \rho \frac{dy'}{ds'} \text{.......................} (2).$$

If then s' is known in terms of x', or of y', or of both, by means of this relation and the known equation to the curve we may find $\frac{dx'}{ds'}$ and $\frac{dy'}{ds'}$; and ρ is known from the equation $s' \mp \rho = l$. It only remains then to eliminate x' and y' from (1) and (2) and the known equation to the curve; we obtain thus an equation between x and y, which is the required equation to the involute.

100. *Application to the Catenary.*

The equation to the catenary is

$$y' = \frac{c}{2}\left(e^{\frac{x'}{c}} + e^{-\frac{x'}{c}}\right),$$

and $$s' = \frac{c}{2}\left(e^{\frac{x'}{c}} - e^{-\frac{x'}{c}}\right),$$

supposing s' measured from the point for which $x' = 0$ and $y' = c$; we shall now find the equation to that involute to the catenary which begins at the point of the curve just specified.

We have then

$$\frac{dy'}{dx'} = \frac{s'}{c}, \quad \frac{ds'}{dx'} = \frac{y'}{c};$$

thus $$\frac{dy'}{ds'} = \frac{s'}{y'}, \quad \frac{dx'}{ds'} = \frac{c}{y'};$$

and $\rho = s'$, no constant being required, because by supposition ρ vanishes with s'.

Hence equations (1) and (2) of the preceding article become

$$x = x' - \frac{s'c}{y'};$$

$$y = y' - \frac{s'^2}{y'} = \frac{y'^2 - s'^2}{y'} = \frac{c^2}{y'}.$$

And $$s' = \sqrt{(y'^2 - c^2)} = \sqrt{\left(\frac{c^4}{y^2} - c^2\right)} = \frac{c}{y}\sqrt{(c^2 - y^2)};$$

therefore $$\frac{s'}{y'} = \frac{\sqrt{(c^2 - y^2)}}{c};$$

thus $x = x' - \sqrt{(c^2 - y^2)}$; therefore $x' = \sqrt{(c^2 - y^2)} + x$.

We have then to substitute these values of x' and y' in the equation to the catenary, and thus obtain the required relation between x and y. The substitution may be conveniently performed thus,

$$y' = \frac{c}{2}\left(e^{\frac{x'}{c}} + e^{-\frac{x'}{c}}\right);$$

therefore $$\sqrt{(y'^2 - c^2)} = \frac{c}{2}\left(e^{\frac{x'}{c}} - e^{-\frac{x'}{c}}\right);$$

therefore $$y' + \sqrt{(y'^2 - c^2)} = ce^{\frac{x'}{c}},$$

therefore $$x' = c \log \frac{y' + \sqrt{(y'^2 - c^2)}}{c}.$$

Thus finally, $$x + \sqrt{(c^2 - y^2)} = c \log \frac{c + \sqrt{(c^2 - y^2)}}{y}.$$

This curve is called the *tractory;* on account of the radical, there are two values of x for every value of y less than c, these two values being numerically equal, but of opposite signs. There is a cusp at the point for which $x = 0$ and $y = c$; and the axis of x is an asymptote.

101. The polar formulæ may also be used in like manner to determine the involute when the length of an arc of the evolute can be expressed in terms of the polar co-ordinates of its variable extremity. We have (*Diff. Cal.* Art. 332)

$$r'^2 = \rho^2 + r^2 - 2\rho p \quad \text{.....................} (1),$$

$$p'^2 = r^2 - p^2 \quad \text{..........................} (2).$$

Here, as before, the accented letters belong to the known curve, that is, to the evolute, and the unaccented letters to the required involute; thus since the evolute is known, there is a known relation between p' and r'. And $s' \mp \rho = l$, so that if s' can be expressed in terms of p' and r' we may eliminate p' and r' by means of (1), (2), and the known relation between p' and r'. Thus we obtain an equation connecting p and r, which serves to determine the involute.

102. *Application to the Equiangular Spiral.*

In this curve $p' = r' \sin \alpha$, where α is the constant angle of the spiral. If we suppose the involute to begin from the pole of the spiral, and s' to be measured from that point, we have $\rho = s' = r' \sec \alpha$ (Art. 84). Thus (1) of the preceding article becomes

$$r'^2 = r'^2 \sec^2 \alpha + r^2 - 2r'p \sec \alpha$$

$$= r'^2 \sec^2 \alpha + r'^2 \sin^2 \alpha + p^2 - 2r'p \sec \alpha, \text{ by (2)}.$$

From this quadratic for p we obtain

$$p - r' \sec \alpha = \pm r' \cos \alpha.$$

If we take the upper sign we find $p = \dfrac{r'(1 + \cos^2 \alpha)}{\cos \alpha}$, and then from (2) we find $r^2 = \dfrac{1 + 3\cos^2 \alpha}{\cos^2 \alpha} r'^2$. But this solution must be rejected, because from it we should find ρ or $r\dfrac{dr}{dp} = \dfrac{1 + 3\cos^2 \alpha}{\cos \alpha (1 + \cos^2 \alpha)} r'$, which is inconsistent with the equation $\rho = r' \sec \alpha$.

If we take the lower sign we find $p = \dfrac{r' \sin^2 \alpha}{\cos \alpha}$, and then from (2) we find $r^2 = \dfrac{r'^2 \sin^2 \alpha}{\cos^2 \alpha}$; thus $p = r \sin \alpha$. Hence the involute is an equiangular spiral with the same constant angle as the evolute has.

Intrinsic Equation to a Curve.

103. Let s denote the length of an arc of a curve measured from some fixed point, ϕ the inclination of the tangent at the variable extremity to the tangent at some fixed point of the curve; then the equation which determines the relation between s and ϕ is called the *intrinsic equation* to the curve. In some investigations, especially those relating to involutes and evolutes, this method of determining a curve is simpler than the ordinary method of referring the curve to rectangular axes which are *extrinsic* lines.

104. We will first shew how the *intrinsic* equation may be obtained from the ordinary equation.

Suppose $y = f(x)$ the equation to a curve, the origin being a point on the curve, and the axis of y a tangent at that point; from the given equation we have

$$\frac{dy}{dx} = f'(x) = \frac{1}{\tan\phi} \text{ by hypothesis};$$

thus x is known in terms of $\tan\phi$, say $x = F(\tan\phi)$; then

$$\frac{dx}{d\phi} = F'(\tan\phi)\sec^2\phi;$$

also

$$\frac{ds}{dx} = \operatorname{cosec}\phi;$$

therefore

$$\frac{ds}{d\phi} = F'(\tan\phi)\sec^2\phi\operatorname{cosec}\phi;$$

from this equation s may be found in terms of ϕ by integration. A similar result will be obtained if at the origin the axis of x be the axis which we suppose to coincide with a tangent.

105. *Application to the Cycloid.*

By *Diff. Cal.* Art. 358, we have

$$\frac{dy}{dx} = \sqrt{\left(\frac{2a - x}{x}\right)} = \frac{1}{\tan\phi};$$

therefore $$\frac{2a}{x} = \frac{1}{\sin^2\phi}, \quad x = 2a\sin^2\phi,$$

$$\frac{dx}{d\phi} = 4a\sin\phi\cos\phi,$$

$$\frac{ds}{d\phi} = \operatorname{cosec}\phi\,\frac{dx}{d\phi} = 4a\cos\phi;$$

therefore $$s = 4a\sin\phi + C.$$

The constant will be zero if we suppose s measured from the fixed point where the first tangent is drawn, that is, from the vertex of the curve.

106. Having given the intrinsic equation to deduce the ordinary equation.

We have $$\frac{dx}{ds} = \sin\phi;$$

therefore $$x = \int ds \sin\phi.$$

Similarly $$y = \int ds \cos\phi.$$

Now s is by supposition known in terms of ϕ; thus by integration we may find x and y in terms of ϕ, and then by eliminating ϕ we obtain the ordinary equation to the curve in terms of x and y.

107. *Application to the Cycloid.*

Here $s = 4a\sin\phi$;

thus $$x = \int ds \sin\phi = 4a\int \sin\phi\cos\phi\,d\phi = C - a\cos 2\phi,$$

$$y = \int ds \cos\phi = 4a\int \cos^2\phi\,d\phi = C' + 2a\phi + a\sin 2\phi.$$

Hence by eliminating ϕ we can obtain the ordinary equation; if the origin of the rectangular axes is the vertex of the curve, we shall have $C = a$ and $C' = 0$.

108. We shall now give some miscellaneous examples of intrinsic equations.

The intrinsic equation to the circle is obviously $s = a\phi$.

109. The equation to the catenary is

$$y + c = \frac{c}{2}\left(e^{\frac{x}{c}} + e^{-\frac{x}{c}}\right),$$

the origin being on the curve. Hence

$$\frac{dy}{dx} = \tfrac{1}{2}\left(e^{\frac{x}{c}} - e^{-\frac{x}{c}}\right), \quad s = \frac{c}{2}\left(e^{\frac{x}{c}} - e^{-\frac{x}{c}}\right);$$

thus if ϕ be the angle which the tangent at any point makes with the tangent at the origin,

$$s = c \tan \phi.$$

110. We have seen in Art. 86, that for the epicycloid

$$\frac{dy}{dx} = \frac{\cos\theta - \cos\dfrac{a+b}{b}\theta}{\sin\dfrac{a+b}{b}\theta - \sin\theta} = \tan\phi \text{ suppose},$$

thus
$$\phi = \frac{a+2b}{2b}\theta.$$

Again, from the same article,

$$s = -\frac{\sqrt{(c^2 - a^2)}}{a}\sqrt{(c^2 - r^2)} + C$$

$$= -\frac{4b(a+b)}{a}\cos\frac{a\theta}{2b} + C$$

$$= \frac{4b(a+b)}{a}\left(1 - \cos\frac{a\theta}{2b}\right),$$

if we suppose s measured from the point for which $\theta = 0$.

Thus
$$s = \frac{4b(a+b)}{a}\left(1 - \cos\frac{a\phi}{a+2b}\right).$$

We may simplify this result by putting

$$\phi = \frac{\pi(a+2b)}{2a} + \phi', \quad \text{and } s = \frac{4b(a+b)}{a} + s';$$

this amounts to measuring the arc from a vertex instead of from a cusp. Thus

$$s' = \frac{4b(a+b)}{a} \sin \frac{a\phi'}{a+2b},$$

where the accent may now be dropped.

111. Similarly the intrinsic equation to the hypocycloid may be written

$$s = \frac{4b(a-b)}{a} \sin \frac{a\phi}{a-2b}.$$

112. It appears from the last two articles that $s = l \sin n\phi$ represents an epicycloid or hypocycloid, according as n is less or greater than unity. For example, if

$$s = l \sin \frac{\phi}{2}, \quad s = l \sin \frac{\phi}{3}, \quad s = l \sin \frac{\phi}{4}, \quad s = l \sin \frac{\phi}{5}, \ldots$$

we have epicycloids in which $\frac{b}{a} = \frac{1}{2}, 1, \frac{3}{2}, 2, \ldots$

If $s = l \sin 2\phi, \; s = l \sin 3\phi, \; s = l \sin 4\phi, \; s = l \sin 5\phi, \ldots$

we have hypocycloids in which $\frac{b}{a} = \frac{1}{4}, \frac{1}{3}, \frac{3}{8}, \frac{2}{5}, \ldots$

113. If ρ be the radius of curvature of the curve at the point determined by s and ϕ, we have (*Diff. Cal.* Art. 324)

$$\rho = \frac{ds}{d\phi}.$$

In the logarithmic spiral we know that ρ varies as s if the arc be measured from the pole; thus

$$\rho = ks = \frac{ds}{d\phi};$$

therefore $k = \frac{1}{s}\frac{ds}{d\phi}$, and therefore by integration

$$k\phi + \text{constant} = \log s;$$

therefore $$s = ae^{k\phi},$$

where a is a constant. If we put $s = s' + a$ we have

$$s' = a\,(e^{k\phi} - 1),$$

and now s' is measured from the point for which $\phi = 0$.

114. If the intrinsic equation to a curve be known, that to the evolute can be found.

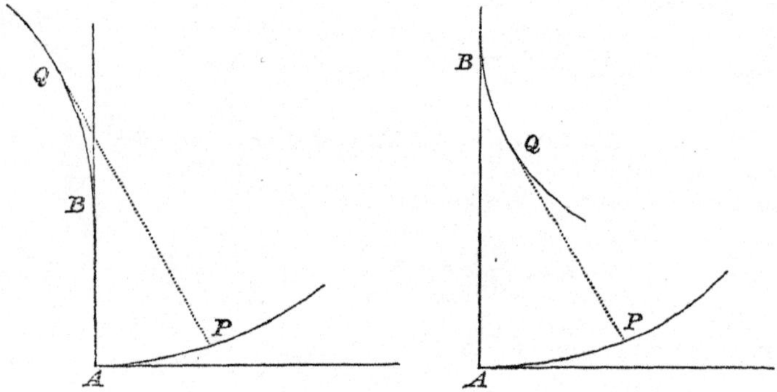

Let AP be a curve, BQ the evolute; let s be the length of an arc of AP measured from some fixed point up to P; s' the length of an arc of BQ measured from some fixed point up to Q. It is evident that ϕ is the same both for s and s', if in BQ we measure ϕ from BA, which is perpendicular to the line from which ϕ is measured in AP.

In the left-hand figure $s' = \rho - C = \dfrac{ds}{d\phi} - C$.

In the right-hand figure $s' = C - \rho = C - \dfrac{ds}{d\phi}$.

Thus if s be known in terms of ϕ, we can find s' in terms of ϕ. The constant C is equal to the value of ρ at the point corresponding to that for which $s' = 0$.

115. For example, in the cycloid $s = 4a \sin \phi$; thus

$$s' = C - 4a \cos \phi.$$

Put $\phi = \psi + \frac{\pi}{2}$ and $s' = \sigma + C$; thus

$$\sigma = 4a \sin \psi.$$

This shews that the evolute is an equal cycloid.

116. Similarly if the intrinsic equation to a curve be known, that to the involute may be found. For by Art. 114

$$\frac{ds}{d\phi} = C \pm s';$$

therefore $$s = \int (C \pm s')\, d\phi.$$

Thus if s' be known in terms of ϕ, we can find s in terms of ϕ.

117. For example, in the circle $s' = a\phi$. Thus

$$s = \int (C \pm a\phi)\, d\phi = C\phi \pm \frac{a\phi^2}{2} + C'.$$

If we suppose s to begin where $\phi = 0$ we have $C' = 0$, and further, if s begins where the involute meets the circle we have $C = 0$; thus $s = \frac{a\phi^2}{2}$. (See *Diff. Cal.* Art. 333.)

118. It is obvious that by the methods of Arts. 114 and 116 we may find the evolute of the evolute of a curve, or the involute of the involute of a curve, and so on.

119. The student may exercise himself in tracing curves from their intrinsic equations; he will find it useful to take such a curve as the cycloid, the form of which is well known, and ascertain that the intrinsic equation does lead to that form; he may then take some of the epicycloids or hypocycloids given in Art. 112. For further information on this subject, and for illustrative figures, the student is referred to two memoirs by Dr Whewell, published in the *Cambridge Philosophical Transactions*, Vol. VIII. page 659, and Vol. IX. page 150.

Curves of double Curvature.

120. Let x, y, z be the co-ordinates of a point on a curve in space; $x+\Delta x$, $y+\Delta y$, $z+\Delta z$ the co-ordinates of an adjacent point on the curve. Then it is known by the principles of solid geometry, that the length of the chord joining these two points is $\sqrt{\{(\Delta x)^2+(\Delta y)^2+(\Delta z)^2\}}$. Let s be the length of the arc of the curve measured from some fixed point up to (x, y, z); and let $s+\Delta s$ be the length of the arc measured from the same fixed point up to $(x+\Delta x, y+\Delta y, z+\Delta z)$. We shall assume that Δs bears to the chord joining the adjacent points a ratio which is ultimately equal to unity when the second point moves along the curve up to the first point. Thus the limit of

$$\frac{\Delta s}{\sqrt{\{(\Delta x)^2+(\Delta y)^2+(\Delta z)^2\}}}, \text{ that is, of } \frac{\dfrac{\Delta s}{\Delta x}}{\sqrt{\left\{1+\left(\dfrac{\Delta y}{\Delta x}\right)^2+\left(\dfrac{\Delta z}{\Delta x}\right)^2\right\}}},$$

is unity. Hence

$$\frac{ds}{dx}=\sqrt{\left\{1+\left(\frac{dy}{dx}\right)^2+\left(\frac{dz}{dx}\right)^2\right\}},$$

therefore $$s=\int\sqrt{\left\{1+\left(\frac{dy}{dx}\right)^2+\left(\frac{dz}{dx}\right)^2\right\}}\,dx.$$

From the equations to the curve $\dfrac{dy}{dx}$ and $\dfrac{dz}{dx}$ may be expressed in terms of x, and then by integration s is known in terms of x.

121. With respect to the assumption in the preceding article, the student may refer to *Diff. Cal.* Arts. 307, 308; he may also hereafter consult De Morgan's *Differential and Integral Calculus*, page 444, and Homersham Cox's *Integral Calculus*, page 95.

122. Suppose, for example, that the curve is determined by the equations

$$y^2 = 4ax \;\ldots\ldots\ldots\ldots\ldots\ldots\ldots\ldots(1),$$

$$z=\sqrt{(2cx-x^2)}+c\,\text{vers}^{-1}\frac{x}{c}\ldots\ldots\ldots(2),$$

so that the curve is formed by the intersection of two cylinders, namely a cylinder which has its generating lines parallel to the axis of z, and which stands upon the parabola in the plane of (x, y) given by (1), and a cylinder which has its generating lines parallel to the axis of y, and which stands on the cycloid in the plane of (x, z) given by (2). Then

$$\frac{dy}{dx} = \sqrt{\left(\frac{a}{x}\right)}, \quad \frac{dz}{dx} = \sqrt{\left(\frac{2c - x}{x}\right)};$$

hence
$$\frac{ds}{dx} = \sqrt{\left(1 + \frac{a}{x} + \frac{2c}{x} - 1\right)} = \sqrt{\left(\frac{2c + a}{x}\right)};$$

therefore
$$s = \sqrt{(2c + a)} \int \frac{dx}{\sqrt{x}} = 2\sqrt{(2c + a)}\sqrt{x}.$$

No constant is required if we measure the arc from the origin of co-ordinates.

123. The formula given in Art. 120 may be changed into

$$s = \int \sqrt{\left\{1 + \left(\frac{dx}{dy}\right)^2 + \left(\frac{dz}{dy}\right)^2\right\}}\, dy,$$

and
$$s = \int \sqrt{\left\{1 + \left(\frac{dx}{dz}\right)^2 + \left(\frac{dy}{dz}\right)^2\right\}}\, dz,$$

and in some cases these forms may be more convenient than that in Art. 120.

124. Sometimes a curve in space is determined by three equations, which express x, y, z respectively in terms of an auxiliary variable; then by eliminating this variable, we may, if necessary, obtain two equations connecting x, y, and z, and thus determine the curve in the ordinary way. Suppose then x, y, z each a known function of t; then

$$\frac{dy}{dx} = \frac{\dfrac{dy}{dt}}{\dfrac{dx}{dt}}, \quad \text{and} \quad \frac{dz}{dx} = \frac{\dfrac{dz}{dt}}{\dfrac{dx}{dt}};$$

and $$s = \int \sqrt{\left\{1 + \left(\frac{dy}{dx}\right)^2 + \left(\frac{dz}{dx}\right)^2\right\}} \, dx$$

$$= \int \sqrt{\left\{1 + \left(\frac{dy}{dx}\right)^2 + \left(\frac{dz}{dx}\right)^2\right\}} \frac{dx}{dt} \, dt$$

$$= \int \sqrt{\left\{\left(\frac{dx}{dt}\right)^2 + \left(\frac{dy}{dt}\right)^2 + \left(\frac{dz}{dt}\right)^2\right\}} \, dt.$$

125. *Application to the Helix.*

This curve may be determined by the equations

$$x = a \cos t, \qquad y = a \sin t, \qquad z = ct;$$

thus $$s = \sqrt{(a^2 + c^2)} \int dt = t \sqrt{(a^2 + c^2)} + C.$$

126. When polar co-ordinates are used to determine the position of a point in space, we have the following equations connecting the rectangular and polar co-ordinates of any point,

$$x = r \sin \theta \cos \phi, \qquad y = r \sin \theta \sin \phi, \qquad z = r \cos \theta.$$

And as a curve in space is determined by two equations between x, y, and z, it may also be determined by two equations between r, θ, and ϕ. Thus we may conceive r and ϕ to be known functions of θ, and therefore x, y, and z become known functions of θ.

Hence

$$\frac{dx}{d\theta} = \sin \theta \cos \phi \frac{dr}{d\theta} - r \sin \theta \sin \phi \frac{d\phi}{d\theta} + r \cos \theta \cos \phi,$$

$$\frac{dy}{d\theta} = \sin \theta \sin \phi \frac{dr}{d\theta} + r \sin \theta \cos \phi \frac{d\phi}{d\theta} + r \cos \theta \sin \phi,$$

$$\frac{dz}{d\theta} = \cos \theta \frac{dr}{d\theta} - r \sin \theta.$$

Therefore $$\left(\frac{dx}{d\theta}\right)^2 + \left(\frac{dy}{d\theta}\right)^2 + \left(\frac{dz}{d\theta}\right)^2 = \left(\frac{dr}{d\theta}\right)^2 + r^2 \sin^2 \theta \left(\frac{d\phi}{d\theta}\right)^2 + r^2,$$

and $$s = \int \sqrt{\left\{r^2 + \left(\frac{dr}{d\theta}\right)^2 + r^2 \sin^2\theta \left(\frac{d\phi}{d\theta}\right)^2\right\}}\, d\theta.$$

This may be transformed into

$$s = \int \sqrt{\left\{r^2 \left(\frac{d\theta}{dr}\right)^2 + 1 + r^2 \sin^2\theta \left(\frac{d\phi}{dr}\right)^2\right\}}\, dr$$

or into $$s = \int \sqrt{\left\{r^2 \left(\frac{d\theta}{d\phi}\right)^2 + \left(\frac{dr}{d\phi}\right)^2 + r^2 \sin^2\theta\right\}}\, d\phi.$$

127. If p be the perpendicular from the origin on the tangent to a curve in space, then the equation

$$\frac{ds}{dr} = \frac{r}{\sqrt{(r^2 - p^2)}},$$

which was proved for a *plane* curve in Art. 85, will still hold. For each member of the equation expresses the secant of the angle which the tangent makes with the radius vector at the point of contact.

Therefore $$s = \int \frac{r\,dr}{\sqrt{(r^2 - p^2)}}.$$

EXAMPLES.

1. For what values of m and n are the curves $a^m y^n = x^{m+n}$ *rectifiable?* (See Art. 15.)

Result. If $\frac{n}{2m}$ or $\frac{n}{2m} + \frac{1}{2}$ is an integer,

2. Shew that the length of the arc of a Tractory measured from the cusp is determined by $s = c \log \frac{c}{y}$.

3. Shew that the Cissoid is rectifiable.

4. Shew that the whole length of the curve whose equation is $4(x^2 + y^2) - a^2 = 3a^{\frac{4}{3}} y^{\frac{2}{3}}$ is equal to $6a$.

$\left[\text{It may be proved that } \left(\frac{ds}{dy}\right)^2 = \frac{a^{\frac{4}{3}}}{4y^{\frac{2}{3}}\left(a^{\frac{2}{3}} - y^{\frac{2}{3}}\right)}\right]$.

5. The length of the arc of the curve

$$(x+y)^{\frac{2}{3}} - (x-y)^{\frac{2}{3}} = a^{\frac{2}{3}}$$

between the limits (x_1, y_1) and (x, y) is

$$\frac{1}{2\sqrt{2}}\{(x+y)^{\frac{2}{3}} + (x-y)^{\frac{2}{3}}\}^{\frac{3}{2}} - \frac{1}{2\sqrt{2}}\{(x_1+y_1)^{\frac{2}{3}} + (x_1-y_1)^{\frac{2}{3}}\}^{\frac{3}{2}}.$$

6. If $s = ae^{\frac{x}{c}}$, find the relation between x and y.

7. Shew that the intrinsic equation to the parabola is

$$\frac{ds}{d\phi} = \frac{2a}{\cos^3\phi} \text{ or } s = \frac{a}{2}\log\frac{1+\sin\phi}{1-\sin\phi} + \frac{a\sin\phi}{1-\sin^2\phi}.$$

8. The intrinsic equation to the curve $y^3 = ax^2$ is

$$s = \frac{8a}{27}(\sec^3\phi - 1).$$

9. Shew that the length of the arc of the evolute of a parabola from the cusp to the point where the evolute meets the parabola is $2a(3\sqrt{3}-1)$; where $4a$ is the latus rectum of the parabola.

10. The evolute of an epicycloid is an epicycloid, the radius of the fixed circle being $\frac{a^2}{a+2b}$ and the radius of the generating circle $\frac{ab}{a+2b}$. (Arts. 110 and 114.)

11. Shew that if the equation to a curve be found by eliminating θ between the equations

$$x = \sin\theta\,\psi'(\theta) + \cos\theta\,\psi''(\theta),$$

and $$y = \cos\theta\,\psi'(\theta) - \sin\theta\,\psi''(\theta),$$

then $$s = \psi(\theta) + \psi''(\theta).$$

12. Shew that the length of the curve $8a^3y = x^4 + 6a^2x^2$ measured from the origin is $\frac{x}{8a^3}(x^2+4a^2)^{\frac{3}{2}}$.

CHAPTER VII.

AREAS OF PLANE CURVES AND OF SURFACES.

Plane Areas. Rectangular Formulæ. Single Integration.

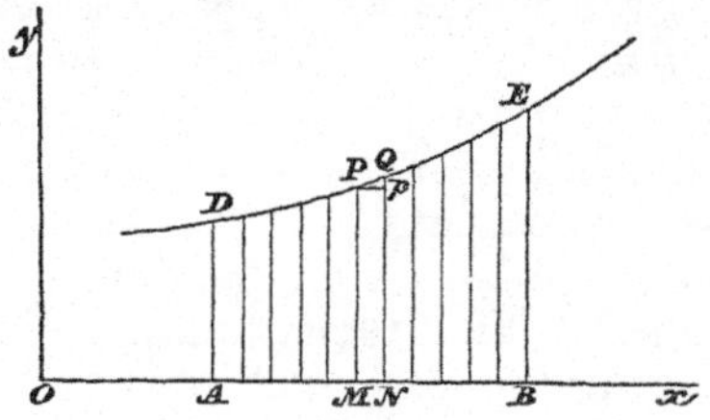

128. Let DPE be a curve, of which the equation is $y = \phi(x)$, and suppose x, y to be the co-ordinates of a point P. Let A denote the area included between the curve, the axis of x, the ordinate PM, and some fixed ordinate AD, then (*Diff. Cal.* Art. 43)

$$\frac{dA}{dx} = \phi(x);$$

hence
$$A = \int \phi(x)\, dx.$$

Let $\psi(x) + C$ be the integral of $\phi(x)$; thus

$$A = \psi(x) + C.$$

Let A_1 denote the area when the variable ordinate is at a distance x from the axis of y, and let A_2 denote the area when the variable ordinate is at a distance x_2 from the axis of y; then

$$A_1 = \psi(x_1) + C, \qquad A_2 = \psi(x_2) + C;$$

therefore $$A_2 - A_1 = \psi(x_2) - \psi(x_1) = \int_{x_1}^{x_2} \phi(x)\, dx.$$

129. *Application to the Circle.*

The equation to the circle referred to its centre as origin is $y^2 = a^2 - x^2$; here $\phi(x) = \sqrt{(a^2 - x^2)}$; thus

$$A = \int \phi(x)\, dx = \int \sqrt{(a^2 - x^2)}\, dx = \frac{x\sqrt{(a^2 - x^2)}}{2} + \frac{a^2}{2} \sin^{-1}\frac{x}{a} + C.$$

The constant C vanishes if we suppose the *fixed ordinate* to coincide with the axis of y. It will be seen by drawing a figure, that the area comprised between the axis of x, the axis of y, the circle, and the ordinate at the distance x from the axis of y, may be divided into a triangle and a sector, the values of which are given by the first and second terms in the above expression for A. This remark may serve to assist the student in remembering the important integral

$$\int \sqrt{(a^2 - x^2)}\, dx = \frac{x\sqrt{(a^2 - x^2)}}{2} + \frac{a^2}{2} \sin^{-1}\frac{x}{a}.$$

130. *Application to the Ellipse.*

Suppose it required to find the whole area of the ellipse. The equation to the ellipse may be written $y^2 = \frac{b^2}{a^2}(a^2 - x^2)$. Hence the area of one quadrant of the ellipse

$$= \int_0^a \frac{b}{a}\sqrt{(a^2 - x^2)}\, dx = \frac{b}{a}\int_0^a \sqrt{(a^2 - x^2)}\, dx = \frac{b}{a}\frac{\pi a^2}{4} = \frac{\pi ab}{4};$$

hence the area of the ellipse is πab.

131. *Application to the Parabola.*

The equation to the parabola is $y^2 = 4ax$; here then

$$\phi(x) = \sqrt{(4ax)},$$

and $$\int \sqrt{(4ax)}\, dx = \frac{4\sqrt{a}}{3} x^{\frac{3}{2}} + C;$$

thus with the notation of Art. 128

$$A_2 - A_1 = \int_{x_1}^{x_2} \sqrt{(4ax)}\, dx = \frac{4\sqrt{a}}{3} (x_2^{\frac{3}{2}} - x_1^{\frac{3}{2}}).$$

If $x_1 = 0$ we have for the area $\frac{4\sqrt{a}}{3} x_2^{\frac{3}{2}}$, that is, two thirds of the product of the abscissa x_2 and the ordinate $\sqrt{(4ax_2)}$.

132. *Application to the Cycloid.*

The integration required by the formula $\int y\,dx$ becomes sometimes more easy if we express x and y in terms of a new variable. Thus, for example, in the cycloid we can put (*Diff. Cal.* Art. 358)

$$x = a(1 - \cos\theta), \qquad y = a(\theta + \sin\theta);$$

therefore $$\int y\,dx = a^2 \int (\theta + \sin\theta) \sin\theta\, d\theta$$

$$= a^2 \int \theta \sin\theta\, d\theta + \frac{a^2}{2} \int (1 - \cos 2\theta)\, d\theta;$$

this gives $$a^2(-\theta\cos\theta + \sin\theta) + \frac{a^2}{2}\left(\theta - \frac{\sin 2\theta}{2}\right).$$

If we take this between the limits 0 and π for θ, we obtain the area of half a cycloid; the result is $\frac{3a^2\pi}{2}$. Hence the area of the whole cycloid is equal to three times that of the generating circle.

133. The equations to the *companion to the cycloid* are

$$x = a(1 - \cos\theta), \quad y = a\theta;$$

hence it may be shewn that the area of the whole curve is twice that of the generating circle.

134. If a curve be determined by the equation $x = \phi(y)$, then the area contained between the curve, the axis of y, and

lines drawn parallel to the axis of x at distances respectively equal to y_1 and y_2 is $\int_{y_1}^{y_2} \phi(y)\, dy$. This is obvious after the proof of the similar proposition in Art. 128.

135. The formulæ in Arts. 128 and 134 furnish one of the most simple and important examples of the application of the Integral Calculus. As we have already remarked, the problem of determining the areas of curves was one of those which gave rise to the Integral Calculus, and the symbols used are very expressive of the process necessary for solving the problem. In the figure to Art. 128, the student will see that the rectangle $PpNM$ may be appropriately denoted by $y\Delta x$, and the process of finding the area of $ADEB$ amounts to this; we first effect the addition denoted by $\Sigma y \Delta x$, and then diminish Δx indefinitely.

136. Suppose we require the area contained between the curve $y = c \sin \frac{x}{a}$, the axis of x, and ordinates at the distances x_1 and x_2 respectively from the axis of y. We have

$$c \int_{x_1}^{x_2} \sin \frac{x}{a}\, dx = ca \left(\cos \frac{x_1}{a} - \cos \frac{x_2}{a} \right).$$

Suppose then $x_1 = 0$ and $x_2 = a\pi$; the area is $2ca$. Next suppose $x_1 = 0$ and $x_2 = 2a\pi$; the result

$$ca \left(\cos \frac{x_1}{a} - \cos \frac{x_2}{a} \right)$$

becomes zero in this case, which is obviously inadmissible, since the area must be some positive quantity. In fact $\sin \frac{x}{a}$ is *negative* from $x = a\pi$ to $x = 2a\pi$, but in the proof that the area is equal to $\int y\,dx$, it is supposed that y is *positive*. If y be really negative the area will be $\int (-y)\, dx$.

Thus in the present example the area will not be

$$\int_0^{2a\pi} \sin \frac{x}{a}\, dx \quad \text{but} \quad c \int_0^{a\pi} \sin \frac{x}{a}\, dx + c \int_{a\pi}^{2a\pi} \left(- \sin \frac{x}{a} \right) dx,$$

that is, $$c\int_0^{a\pi} \sin\frac{x}{a}\,dx - c\int_{a\pi}^{2a\pi} \sin\frac{x}{a}\,dx;$$

this will give $2ca + 2ca$, that is, $4ca$.

Plane Areas. Rectangular Formulæ. Double Integration.

137. In Art. 128 we have obtained a formula for finding the area of a curve; that formula supposes the area to be the limit of a number of elemental areas, each element being a quantity of which $y\Delta x$ is the type. We shall now proceed to explain another mode of decomposing the required area into elemental areas.

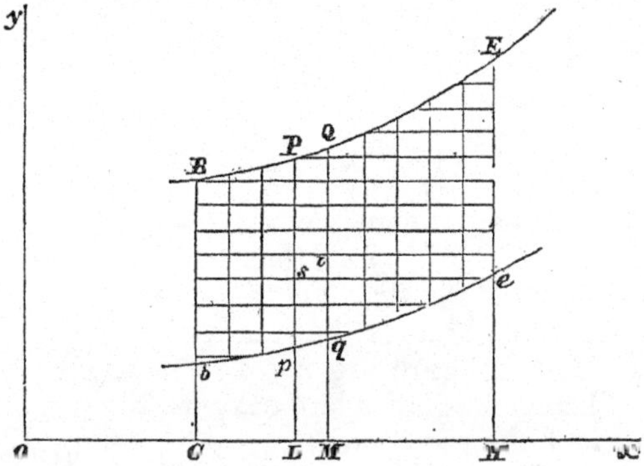

Suppose we require the area included between the curves $BPQE$ and $bpqe$, and the straight lines Bb and Ee. Let a series of lines be drawn parallel to the axis of y, and another series parallel to the axis of x. Let st represent one of the rectangles thus formed, and suppose x and y to be the co-ordinates of s, and $x + \Delta x$ and $y + \Delta y$ the co-ordinates of t; then the area of the rectangle st is $\Delta x \Delta y$. Hence the required area may be found by summing up all the values of $\Delta x \Delta y$, and then proceeding to the limit obtained by supposing Δx and Δy to diminish indefinitely.

We effect the required summation of such terms as $\Delta x \Delta y$ in the following way: we first collect all the rectangles similar to st which are contained in the strip $PQqp$, and we thus obtain the area of this strip; then we sum up all the strips similar to this strip which lie between Bb and Ee. The error we may make by neglecting the element of area which lies at the top and bottom of each strip, and which is not a complete rectangle, will disappear in the limit when Δx and Δy are indefinitely diminished.

Let $y = \phi(x)$ be the equation to the upper curve, and $y = \psi(x)$ the equation to the lower curve; let $OC = c$ and $OH = h$, then if A denote the required area, we have

$$A = \int_c^h \int_{\psi(x)}^{\phi(x)} dx\, dy;$$

for the symbolical expression here given denotes the process which we have just stated in words.

Now $\int dy = y$, therefore $\int_{\psi(x)}^{\phi(x)} dy = \phi(x) - \psi(x)$; thus we have

$$A = \int_c^h \{\phi(x) - \psi(x)\}\, dx.$$

In this form we can at once see the truth of the expression, for $\phi(x) - \psi(x) = PL - pL = Pp$; thus $\{\phi(x) - \psi(x)\}\Delta x$ may be taken for the area of the strip $PQqp$, and the formula asserts that A is equal to the limit of the sum of such strips.

The lines in the figure are not necessarily equidistant: that is, the elements of which $\Delta x \Delta y$ is the type are not necessarily all of the same area.

138. The result of the preceding article is, that the area A is found from the equation

$$A = \int_c^h \{\phi(x) - \psi(x)\}\, dx.$$

This result may be obtained in a very simple manner as shewn in the latter part of the preceding article, so that it was not absolutely *necessary* to introduce the formula of double integration. We have however drawn attention to the formula

$$A = \int_c^h \int_{\psi(x)}^{\phi(x)} dx\, dy$$

because of the illustration which is here given of the process of double integration; the student may thus find it easier to apply the processes of double integration to those cases where it is absolutely necessary, of which examples will occur hereafter.

139. If the area which is to be evaluated is bounded by the curves $x = \psi(y)$, and $x = \phi(y)$, and straight lines parallel to the axis of x at distances respectively equal to c and h, we have in a similar manner

$$A = \int_c^h \int_{\psi(y)}^{\phi(y)} dy\, dx = \int_c^h \{\phi(y) - \psi(y)\}\, dy.$$

Some examples of the formulæ of Arts. 137 and 139 will now be considered; we shall see that either of these formulæ may be used in an example, though generally one will be more simple than the other.

140. Required the area included between the parabola $y^2 = ax$ and the circle $y^2 = 2ax - x^2$.

The curves pass through the origin and meet at the point for which $x = a$; thus if we take only that area which lies on the positive side of the axis of x, we have

$$A = \int_0^a \{\sqrt{(2ax - x^2)} - \sqrt{(ax)}\}\, dx = \frac{\pi a^2}{4} - \frac{2a^2}{3}.$$

The whole area will therefore be $2\left(\frac{\pi a^2}{4} - \frac{2a^2}{3}\right)$.

Suppose that we wish in this example to integrate with respect to x first. From the equation $y^2 = 2ax - x^2$ we deduce $x = a \pm \sqrt{(a^2 - y^2)}$, and it will appear at once from a figure that we must take the lower sign in the present question. Thus let x_1 stand for $a - \sqrt{(a^2 - y^2)}$, and x_2 for $\frac{y^2}{a}$, then

$$A = \int_0^a \int_{x_1}^{x_2} dy\, dx = \int_0^a \left\{\frac{y^2}{a} - a + \sqrt{(a^2 - y^2)}\right\} dy$$

$$= \frac{a^2}{3} - a^2 + \frac{\pi a^2}{4} = \frac{\pi a^2}{4} - \frac{2a^2}{3}, \text{ as before.}$$

The reader should draw the figure and pay close attention to the *limits* of the integrations.

141. In the accompanying figure S is the centre of a circle BLD, S is also the focus of a parabola ALC; we shall

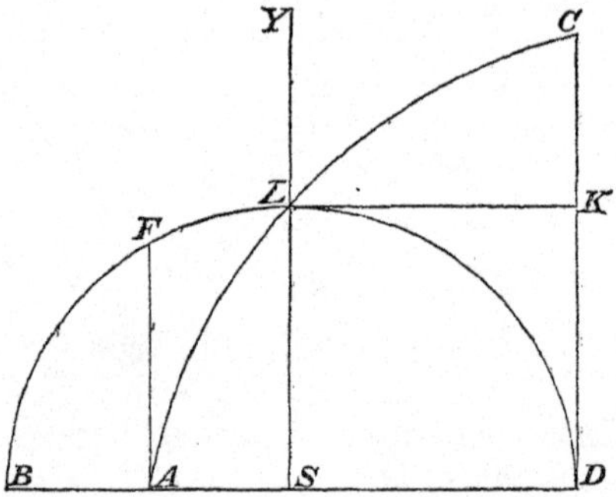

indicate the integrations that should be performed in order to obtain the areas ALB and LDC. This example is introduced for the purpose of illustrating the processes of double integration, and not for any interest in the results: the areas can be easily ascertained by means of formulæ already given; thus ALB is the difference of the parabolic area ALS and the quadrant SLB; and similarly LDC is known.

Take S for origin. In finding the area ALB it will be convenient to suppose the positive direction of the axis of x to be that towards the left hand; thus if $4a$ be the latus rectum of the parabola, and therefore $2a$ the radius of the circle, the equation to the parabola is $y^2 = 4a(a-x)$, and that to the circle $y^2 = 4a^2 - x^2$.

Suppose we integrate with respect to x first, then

$$\text{area } ALB = \int_0^{2a} \int_{x_1}^{x_2} dy\,dx,$$

where $$x_1 = a - \frac{y^2}{4a}, \quad x_2 = \sqrt{(4a^2 - y^2)}.$$

For here $(x_2 - x_1)\,\Delta y$ represents a strip included between the two curves and two lines parallel to the axis of x; and strips are situated at distances from the axis of x ranging between 0 and $2a$, so that the integration with respect to y is taken between the limits 0 and $2a$.

Suppose we integrate with respect to y first; we shall then have to divide the area into two parts by the line AF. Let

$$y_1 = \sqrt{(4a^2 - 4ax)}, \qquad y_2 = \sqrt{(4a^2 - x^2)};$$

then
$$\text{area } ALF = \int_0^a \int_{y_1}^{y_2} dx\, dy = \int_0^a (y_2 - y_1)\, dx;$$

$$\text{area } AFB = \int_a^{2a} \int_0^{y_2} dx\, dy = \int_a^{2a} y_2\, dx;$$

the sum of these two parts expresses the area ALB.

Next take the area LDC; suppose now the positive direction of the axis of x to be that towards the right hand, then the equation to the parabola is $y^2 = 4a(a + x)$, and that to the circle $y^2 = 4a^2 - x^2$.

Suppose we integrate with respect to y first; let

$$y_1 = \sqrt{(4a^2 - x^2)} \text{ and } y_2 = \sqrt{(4a^2 + 4ax)};$$

then
$$\text{area } DLC = \int_0^{2a} \int_{y_1}^{y_2} dx\, dy.$$

Suppose we integrate with respect to x first; we shall then have to divide the area into two parts by the line LK. Let

$$x_1 = \sqrt{(4a^2 - y^2)}, \quad x_2 = \frac{y^2}{4a} - a;$$

then we shall find that $DC = 2a\sqrt{3} = b$ suppose; thus

$$\text{area } DLK = \int_0^{2a} \int_{x_1}^{2a} dy\, dx,$$

$$\text{area } CLK = \int_{2a}^{b} \int_{x_2}^{2a} dy\, dx;$$

the sum of these two parts expresses the area LDC.

142. One case in which the formulæ of Arts. 137 and 139 are useful is that in which the bounding curves are different branches of the same curve. Suppose the equation to a curve to be $(y - mx - c)^2 = a^2 - x^2$; thus

$$y = mx + c \pm \sqrt{(a^2 - x^2)}.$$

Here we may put

$$\psi(x) = mx + c - \sqrt{(a^2 - x^2)},$$

$$\phi(x) = mx + c + \sqrt{(a^2 - x^2)};$$

thus $\phi(x) - \psi(x) = 2\sqrt{(a^2 - x^2)}$, and the complete area of the curve is

$$\int_{-a}^{a} 2\sqrt{(a^2 - x^2)}\,dx, \text{ that is, } \pi a^2.$$

143. We have hitherto supposed the axes rectangular, but if they are oblique and inclined at an angle ω, the formula in Art. 128 becomes

$$A = \sin\omega \int \phi(x)\,dx,$$

and a similar change is made in all the other formulæ. It is obvious that such elements of area as are denoted by $y\Delta x$ and $\Delta y \Delta x$ when the axes are rectangular will be denoted by $\sin\omega\, y\Delta x$ and $\sin\omega\, \Delta y\, \Delta x$ when the axes are inclined at an angle ω.

For example, the equation to the parabola is $y^2 = 4a'x$ when the axes are the oblique system formed by a diameter and the tangent at its extremity; hence the area included between the curve, the axis of x, and an ordinate at the point for which $x = c$, is

$$\sin\omega \int_0^c \sqrt{(4a'x)}\,dx = \frac{4 \sin\omega \sqrt{a'}c^{\frac{3}{2}}}{3},$$

that is, two thirds of the parallelogram which has the abscissa c and the ordinate at its extremity for adjacent sides.

Plane Areas. Polar Formulæ. Single Integration.

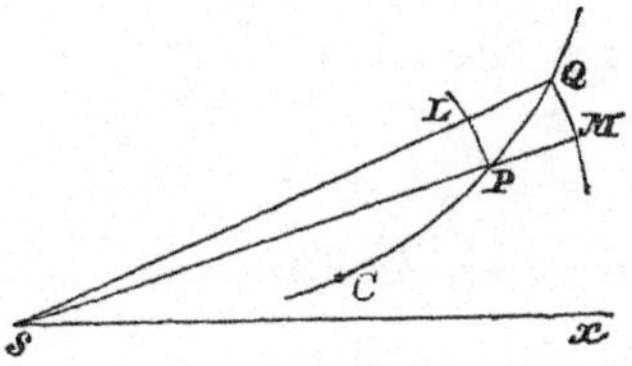

144. Let CPQ be a curve, of which the polar equation is $r=\phi(\theta)$, and suppose r, θ to be the co-ordinates of a point P. Let A denote the area included between the curve, the radius vector SC drawn to a fixed point C, and the radius vector SP, then (*Diff. Cal.* Art. 313)

$$\frac{dA}{d\theta} = \frac{\{\phi(\theta)\}^2}{2}.$$

Hence $$A = \tfrac{1}{2}\int \{\phi(\theta)\}^2 \, d\theta.$$

Let $\psi(\theta)$ be the integral of $\dfrac{\{\phi(\theta)\}^2}{2}$, then

$$A = \psi(\theta) + C.$$

Let A_1 denote the area when the variable radius vector is at an angular distance θ_1 from the initial line, and let A_2 denote the area when the variable radius vector is at an angular distance θ_2 from the initial line; then

$$A_1 = \psi(\theta_1) + C, \qquad A_2 = \psi(\theta_2) + C,$$

therefore $$A_2 - A_1 = \psi(\theta_2) - \psi(\theta_1) = \tfrac{1}{2}\int_{\theta_1}^{\theta_2} \{\phi(\theta)\}^2 \, d\theta.$$

145. *Application to the Equiangular Spiral.*

In this curve $r = be^{\frac{\theta}{c}}$; thus

$$A = \tfrac{1}{2}\int b^2 e^{\frac{2\theta}{c}} \, d\theta = \frac{b^2 c}{4} e^{\frac{2\theta}{c}} + C,$$

and $$A_2 - A_1 = \frac{1}{2}\int_{\theta_1}^{\theta_2} b^2 e^{\frac{2\theta}{c}}\, d\theta = \frac{b^2 c}{4}\left(e^{\frac{2\theta_2}{c}} - e^{\frac{2\theta_1}{c}}\right) = \frac{c}{4}\left(r_2^{\,2} - r_1^{\,2}\right),$$

where r_1 and r_2 are the extreme radii vectores of the area considered.

146. *Application to the Parabola.*

Let the focus be the pole, then

$$r = \frac{a}{\cos^2\frac{\theta}{2}};\ \text{thus } A = \frac{a^2}{2}\int \frac{d\theta}{\cos^4\frac{\theta}{2}}$$

$$= \frac{a^2}{2}\int\left(1 + \tan^2\frac{\theta}{2}\right)\sec^2\frac{\theta}{2}\, d\theta = a^2 \tan\frac{\theta}{2} + \frac{a^2}{3}\tan^3\frac{\theta}{2} + C.$$

Hence $$A_2 - A_1 = a^2\left(\tan\frac{\theta_2}{2} - \tan\frac{\theta_1}{2}\right) + \frac{a^2}{3}\left(\tan^3\frac{\theta_2}{2} - \tan^3\frac{\theta_1}{2}\right).$$

Suppose that $\theta_1 = 0$ and $\theta_2 = \frac{\pi}{2}$, then we obtain for the area $a^2 + \frac{a^2}{3}$, that is, $\frac{4a^2}{3}$; this agrees with Art. 131.

For another example we will suppose the parabola referred to the intersection of the directrix and the axis as pole, the axis being the initial line. Here

$$r = 2a\frac{\cos\theta - \sqrt{(\cos 2\theta)}}{\sin^2\theta},$$

thus $$A = 2a^2\int \frac{\cos^2\theta + \cos 2\theta - 2\cos\theta\sqrt{(\cos 2\theta)}}{\sin^4\theta}\, d\theta$$

$$= 2a^2\int\frac{2\cos^2\theta - \sin^2\theta}{\sin^4\theta}\, d\theta - 4a^2\int\frac{\cos\theta\sqrt{(\cos 2\theta)}}{\sin^4\theta}\, d\theta.$$

Now $$\int \frac{2\cos^2\theta - \sin^2\theta}{\sin^4\theta}\,d\theta = \int (2\cot^2\theta - 1)\,\text{cosec}^2\theta\,d\theta$$
$$= -\tfrac{2}{3}\cot^3\theta + \cot\theta.$$

And $$\int \frac{\cos\theta\,\sqrt{(\cos 2\theta)}\,d\theta}{\sin^4\theta} = \int \frac{\sqrt{(1 - 2\sin^2\theta)}\,d\sin\theta}{\sin^4\theta};$$

assume $\sin\theta = \frac{1}{t}$, then the integral becomes

$$-\int \sqrt{(t^2-2)}\,t\,dt, \text{ that is, } -\tfrac{1}{3}(t^2-2)^{\frac{3}{2}}.$$

Hence, adding the constant, we have

$$A = \frac{4a^2}{3}(\text{cosec}^2\theta - 2)^{\frac{3}{2}} - \frac{4a^2}{3}\cot^3\theta + 2a^2\cot\theta + C$$
$$= 2a^2\cot\theta + \frac{4a^2}{3}\,\frac{(\cos 2\theta)^{\frac{3}{2}} - \cos^3\theta}{\sin^3\theta} + C.$$

The constant will be zero if A commences from the initial line; for it will be found on investigation that

$$2\cot\theta + \frac{4}{3}\,\frac{(\cos 2\theta)^{\frac{3}{2}} - \cos^3\theta}{\sin^3\theta} \text{ vanishes when } \theta = 0.$$

147. *Application to the curve* $r = a(\theta + \sin\theta)$. Here

$$A = \frac{a^2}{2}\int (\theta + \sin\theta)^2\,d\theta = \frac{a^2}{2}\int (\theta^2 + 2\theta\sin\theta + \sin^2\theta)\,d\theta;$$

and $$\int \theta\sin\theta\,d\theta = -\theta\cos\theta + \sin\theta$$

$$\int \sin^2\theta\,d\theta = \tfrac{1}{2}\int (1 - \cos 2\theta)\,d\theta = \frac{\theta}{2} - \frac{\sin 2\theta}{4},$$

thus $$A = \frac{a^2}{2}\left\{\frac{\theta^3}{3} - 2\theta\cos\theta + 2\sin\theta + \frac{\theta}{2} - \frac{\sin 2\theta}{4}\right\} + C.$$

Suppose we require the area of the smallest portion which is bounded by the curve and by a radius vector which is inclined to the initial line at a right angle; then we have 0 and $\frac{1}{2}\pi$ as the limits of the integration. Thus the required area is

$$\frac{a^2}{2}\left\{\frac{\pi^3}{24}+\frac{\pi}{4}+2\right\}.$$

Plane Curves. Polar Formulæ. Double Integration.

148. In Art. 144 we have obtained a formula for finding the area of a curve; that formula supposes the area to be the limit of a number of elemental areas, each element being a quantity of which $\frac{1}{2}r^2\Delta\theta$ is the type. We shall now proceed to explain another mode of decomposing the required area into elemental areas.

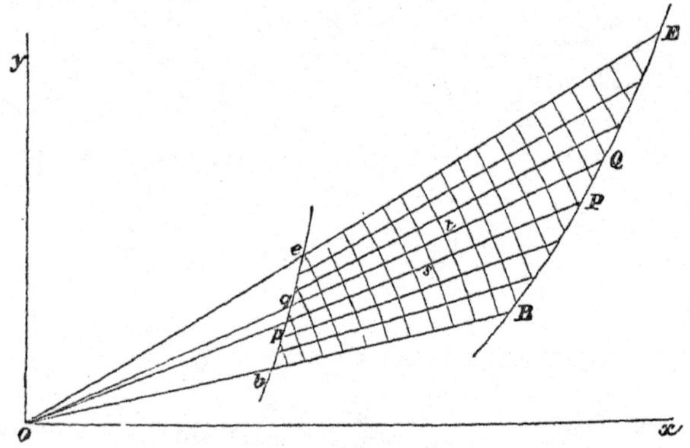

Suppose we require the area included between the curves *BPQE* and *bpqe*, and the straight lines *Bb* and *Ee*. Let a series of radii vectores be drawn from *O*, and a series of circles with *O* as centre; thus the plane area is divided into a series of curvilinear quadrilaterals. Let *st* represent one of these elements, and suppose r and θ to be the polar co-ordinates of s, and $r+\Delta r$ and $\theta+\Delta\theta$ the polar co-ordinates of t; then the area of the element *st* will be ultimately $r\Delta\theta\,\Delta r$. Hence the required area is to be found by summing up all the values of $r\Delta\theta\,\Delta r$, and then proceeding to the limit obtained by supposing $\Delta\theta$ and Δr to diminish indefinitely.

We effect the required summation of such terms as $r\Delta\theta\,\Delta r$ in the following way: we first collect all the elements similar to *st* which are contained in the strip *PQqp*, and thus obtain the area of the strip; then we sum up all the strips similar to this strip which lie between *Bb* and *Ee*.

Let $r=\phi(\theta)$ be the equation to the curve *BPQE* and $r=\psi(\theta)$ the equation to the curve *bpqe*, let α and β be the angles which *OB* and *OE* make respectively with *Ox*; and let *A* denote the required area, then

$$A=\int_\alpha^\beta\int_{\psi(\theta)}^{\phi(\theta)} r\,d\theta\,dr;$$

for the symbolical expression here given denotes the process which we have just stated in words.

Now $\int r\,dr=\frac{r^2}{2}$, therefore

$$\int_{\psi(\theta)}^{\phi(\theta)} r\,dr=\tfrac{1}{2}\left[\{\phi(\theta)\}^2-\{\psi(\theta)\}^2\right],$$

thus we have

$$A=\tfrac{1}{2}\int_\alpha^\beta\left[\{\phi(\theta)\}^2-\{\psi(\theta)\}^2\right]d\theta.$$

In this form we can see at once the truth of the expression, for $OP=\phi(\theta)$ and $Op=\psi(\theta)$, and thus

$$\tfrac{1}{2}\{\phi(\theta)\}^2\Delta\theta-\tfrac{1}{2}\{\psi(\theta)\}^2\Delta\theta$$

may be taken for the area of the strip *PQqp*, and the formula asserts that the area *A* is equal to the limit of the sum of such strips.

149. The remark made in Art. 138 may be repeated here; we have introduced the process in the former part of the preceding article, not because double integration is absolutely necessary for finding the area of a curve, but because the process of finding the area of a curve illustrates double integration.

150. If the area which is to be evaluated is bounded by the curves whose equations are $\theta=\phi(r)$, $\theta=\psi(r)$ respectively,

and by the circles whose equations are $r = a$ and $r = b$ respectively, it will be convenient to integrate with respect to θ first. In this case, instead of first summing up all the elements like *st*, which form the strip *PQqp*, we first sum up all the elements similar to *st* which are included between the two circles which bound *st* and the curves determined by $\theta = \phi(r)$ and $\theta = \psi(r)$. Thus we have

$$A = \int_a^b \int_{\psi(r)}^{\phi(r)} r\,dr\,d\theta.$$

Some examples of the formulæ in Arts. 148 and 150 will now be considered; we shall see that either of these formulæ may be used in an example, although one may be more convenient than the other.

151. We will apply the formulæ to find the area between the two semicircles *OPB* and *Opb* and the straight line *bB*.

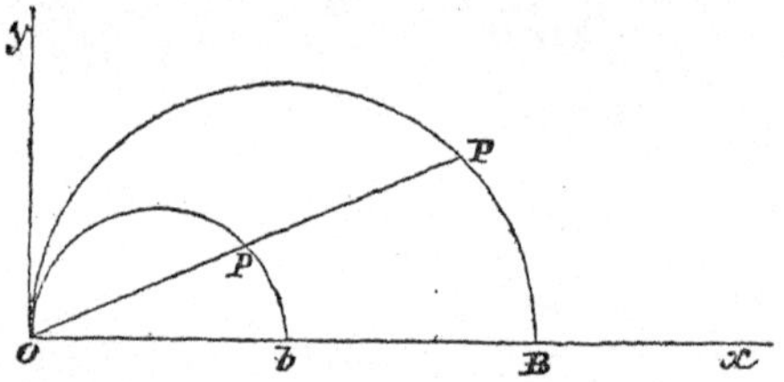

Let $Ob = c$, $OB = h$, then the equation to *OPB* is $r = h\cos\theta$, and the equation to *Opb* is $r = c\cos\theta$. Thus the area

$$= \int_0^{\frac{\pi}{2}} \int_{c\cos\theta}^{h\cos\theta} r\,d\theta\,dr.$$

Now $$\int_{c\cos\theta}^{h\cos\theta} r\,dr = \tfrac{1}{2}(h^2 - c^2)\cos^2\theta;$$

thus the $$\text{area} = \tfrac{1}{2}(h^2 - c^2)\int_0^{\frac{\pi}{2}} \cos^2\theta\,d\theta = \frac{\pi}{8}(h^2 - c^2).$$

Suppose we wish to integrate with respect to θ first; we shall then have to divide the area into two parts by describing an arc of a circle from O as centre, with radius Ob. Then the area bounded by this arc, the straight line Bb, and the larger semicircle is

$$\int_c^h \int_0^{\cos^{-1}\frac{r}{h}} r dr\, d\theta.$$

The area bounded by the aforesaid arc, the semicircle Opb, and the larger semicircle is

$$\int_0^c \int_{\cos^{-1}\frac{r}{c}}^{\cos^{-1}\frac{r}{h}} r dr\, d\theta.$$

The sum of these two parts expresses the required area.

152. Let us apply polar formulæ to the example in Art. 141. With S as pole, the polar equation to the parabola is $r(1+\cos\theta) = 2a$ or $r\cos^2\frac{\theta}{2} = a$, where θ is measured from SB; and the polar equation to the circle is $r = 2a$. Hence, if we integrate with respect to r first,

$$\text{area } ALB = \int_0^{\frac{\pi}{2}} \int_{a\sec^2\frac{\theta}{2}}^{2a} r d\theta\, dr.$$

If we integrate with respect to θ first, we shall have if $\theta_1 = \cos^{-1}\frac{2a-r}{r}$

$$\text{area } ALB = \int_a^{2a} \int_0^{\theta_1} r dr\, d\theta.$$

Next consider the area DLC. The equation to DC is $r\cos\theta = -2a$; the length of SC is $4a$, and the angle BSC is $\frac{2\pi}{3}$. Let $\theta_1 = \cos^{-1}\frac{2a-r}{r}$, $\theta_2 = \cos^{-1}\left(\frac{-2a}{r}\right)$. Then if we integrate with respect to θ first,

$$\text{area } DLC = \int_{2a}^{4a} \int_{\theta_1}^{\theta_2} r dr\, d\theta.$$

If we integrate with respect to r first, we shall have to divide the area into two parts, by the line joining S with C.

The area of the portion which has LC for one of its boundaries is

$$\int_{\frac{\pi}{2}}^{\frac{2\pi}{3}} \int_{2a}^{a\sec^2\frac{\theta}{2}} r d\theta\, dr.$$

The area of the remaining portion is

$$\int_{\frac{2\pi}{3}}^{\pi} \int_{2a}^{-2a\sec\theta} r d\theta\, dr.$$

The sum of these two parts expresses the required area.

153. A good example is supplied by the problem of finding the area included between two radii vectores and two different branches of the same polar curve.

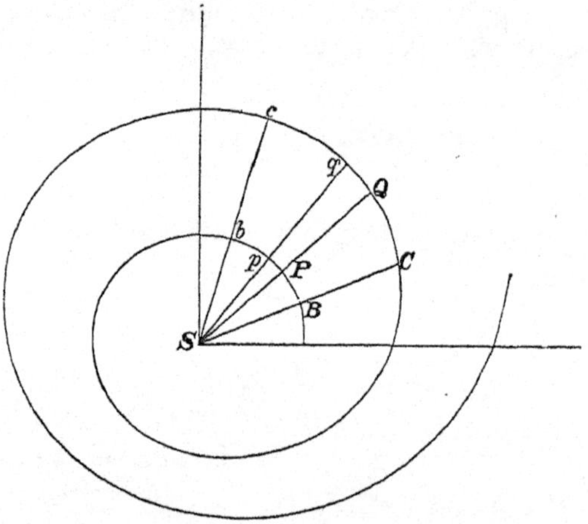

Suppose $BPpb$, $CQqc$ to be two different arcs of a spiral, and that the area is to be evaluated which is bounded by these arcs and the straight lines BC and bc; then the area is

$$\tfrac{1}{2}\int (r_2^2 - r_1^2)\, d\theta,$$

where r_2 denotes any radius vector of the exterior arc, as SQ, and r_1 the corresponding radius vector SP of the interior arc. The limits of θ will be given by the angles which SB and Sb respectively make with the initial line.

Take for example the spiral of Archimedes; let θ be the whole angle which the radius vector has revolved through from the initial line until it takes the position SP; so that θ may be an angle of any magnitude. From the nature of the curve we have SP or $r = a\theta$, where a is some constant. If then CQ is the next branch to BP we shall have

$$SQ = a(\theta + 2\pi).$$

Suppose θ_1 and θ_2 the values of θ for SB and Sb respectively; thus the area $BbcC$

$$= \frac{a^2}{2}\int_{\theta_1}^{\theta_2}\{(\theta + 2\pi)^2 - \theta^2\}\, d\theta$$

$$= \frac{a^2}{2}\{2\pi(\theta_2^2 - \theta_1^2) + 4\pi^2(\theta_2 - \theta_1)\}.$$

154. The student will remark a certain difference between the formulæ $\iint dx\, dy$ and $\iint r\, d\theta\, dr$, which express the area of a plane figure. The former supposes the area decomposed into a number of rectangles and $\Delta x \Delta y$ represents the true area of one rectangle. Hence in taking the aggregate of these rectangles to represent the required area the only error that can arise is owing to the neglect of the irregular elements which occur at the top and bottom of each strip; as we have already remarked in Art. 137. But in the second case $r\Delta\theta\Delta r$ is not the *accurate value* of the area of one of the elements, so that an error is made in the case of every element. It is therefore important to shew formally that the error disappears in the limit, which may be done as follows. The element *st* in the figure of Art. 148 is the difference of two circular sectors, and its exact area is

$$\tfrac{1}{2}(r + \Delta r)^2 \Delta\theta - \tfrac{1}{2} r^2 \Delta\theta,$$

that is, $$r\Delta r\Delta\theta + \tfrac{1}{2}(\Delta r)^2 \Delta\theta.$$

In taking the former term to represent the area we neglect $\frac{1}{2}(\Delta r)^2 \Delta\theta$. Hence the ratio of the term neglected to the term retained

$$=\frac{\frac{1}{2}(\Delta r)^2 \Delta\theta}{r\,\Delta r\,\Delta\theta}=\frac{\Delta r}{2r}.$$

By taking Δr small enough this ratio may be made as small as we please. Hence we may infer that the sum of the neglected terms will ultimately vanish in comparison with the sum of the terms retained, that is, all error disappears in the limit.

Other Polar Formulæ.

155. Let s be the length of the arc of a curve measured from some fixed point up to the point whose co-ordinates are r and θ; let p be the perpendicular from the origin on the tangent at the latter point; then the sine of the angle between this tangent and the corresponding radius vector is $r\frac{d\theta}{ds}$ (*Diff. Cal.* Art. 310); also $\frac{p}{r}$ is another expression for this sine; hence, $r\frac{d\theta}{ds}=\frac{p}{r}$. Let A denote the area between the curve and certain limiting radii vectores; then

$$A=\frac{1}{2}\int r^2 d\theta=\frac{1}{2}\int r^2\frac{d\theta}{ds}ds=\frac{1}{2}\int r\frac{p}{r}ds=\frac{1}{2}\int p\,ds;$$

the limits of s in the latter integral must be such as correspond to the limiting radii vectores of the area considered.

The result can be illustrated geometrically; suppose P, Q adjacent points on a curve, S the pole, p' the perpendicular from S on the chord PQ; then, the area of the triangle PQS

$$=\tfrac{1}{2}p' \times \text{chord } PQ.$$

Now suppose Q to approach indefinitely near to P, then $p'=p$, and the limit of the ratio of the chord PQ to the arc PQ is unity.

156. Since $\int p\,ds = \int p \frac{ds}{dr}\,dr = \int \frac{pr\,dr}{\sqrt{(r^2-p^2)}}$ (Art. 85),

we have
$$A = \tfrac{1}{2}\int \frac{pr\,dr}{\sqrt{(r^2-p^2)}}.$$

157. *Application to the Epicycloid.*

Here $p^2 = \dfrac{c^2(r^2-a^2)}{c^2-a^2}$; thus

$$A = \tfrac{1}{2}\int \frac{c\sqrt{(r^2-a^2)}\,r\,dr}{a\sqrt{(c^2-r^2)}} = \frac{c}{2a}\int \frac{\sqrt{(r^2-a^2)}\,r\,dr}{\sqrt{\{c^2-a^2-(r^2-a^2)\}}}$$

$$= \frac{c}{2a}\int \frac{z^2\,dz}{\sqrt{(c^2-a^2-z^2)}}, \text{ where } z^2 = r^2 - a^2.$$

Now

$$\int \frac{z^2\,dz}{\sqrt{(c^2-a^2-z^2)}} = \int \frac{z^2-(c^2-a^2)}{\sqrt{(c^2-a^2-z^2)}}\,dz + (c^2-a^2)\int \frac{dz}{\sqrt{(c^2-a^2-z^2)}}$$

$$= (c^2-a^2)\int \frac{dz}{\sqrt{(c^2-a^2-z^2)}} - \int \sqrt{(c^2-a^2-z^2)}\,dz$$

$$= \frac{c^2-a^2}{2}\sin^{-1}\frac{z}{\sqrt{(c^2-a^2)}} - \frac{z\sqrt{(c^2-a^2-z^2)}}{2}$$

$$= \frac{c^2-a^2}{2}\sin^{-1}\frac{\sqrt{(r^2-a^2)}}{\sqrt{(c^2-a^2)}} - \frac{\sqrt{(r^2-a^2)}\sqrt{(c^2-r^2)}}{2}.$$

Taking this between the limits $r=a$ and $r=c$, we get $\dfrac{c^2-a^2}{2}\dfrac{\pi}{2}$, that is, $b(a+b)\pi$. Hence the area is $\dfrac{c}{2a}b(a+b)\pi$, that is, $\dfrac{(a+2b)\,b\,(a+b)\,\pi}{2a}$. By doubling this result we obtain the area between the curve and the radii vectores drawn to two consecutive cusps, which is therefore $\dfrac{(a+2b)\,b\,(a+b)\,\pi}{a}$. The area of the circular sector which forms part of this area is πab; subtract the latter and we obtain the area between an arc of the epicycloid extending from one cusp to the next cusp and the fixed circle on which the generating circle rolls; the result is

$$\frac{\pi b^2}{a}(3a+2b).$$

158. Similarly in the hypocycloid the area between the fixed circle and the part of the curve which extends between two consecutive cusps may be found. If a is greater than b the result is

$$\frac{\pi b^2}{a}(3a-2b).$$

Area between a Curve and its Evolute.

159. In the figures to Art. 114, if we suppose the string or line PQ to move through a small angle $\Delta\phi$, the figure between the two positions of the line and the curve AP may be considered ultimately as a sector of a circle; its area will therefore be $\frac{1}{2}\rho^2\Delta\phi$, where $\rho = PQ$. Thus if A denote the whole area bounded by the curve, its evolute, and two radii of curvature corresponding to the values ϕ_1 and ϕ_2 of ϕ, we have

$$A = \frac{1}{2}\int_{\phi_1}^{\phi_2} \rho^2\, d\phi.$$

Since $\frac{d\phi}{ds} = \frac{1}{\rho}$, we may also write this

$$A = \frac{1}{2}\int \rho\, ds,$$

the limits of s being properly taken so as to correspond with the known limits of ϕ.

160. *Application to the Catenary.*

Here $s = c\tan\phi$, Art. 109;

therefore $\rho = c\sec^2\phi, \qquad A = \frac{1}{2}\int_{\phi_1}^{\phi_2} c^2 \sec^4\phi\, d\phi;$

and $$\int \sec^4 \phi \, d\phi = \tan \phi + \tfrac{1}{3} \tan^3 \phi + C;$$

thus A is known.

Area of Surfaces of Revolution. Rectangular Formulæ.

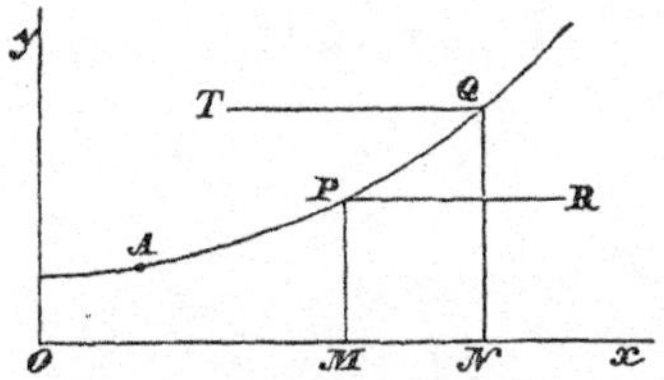

161. Let A be a fixed point in the curve APQ; let x, y be the co-ordinates of any point P, and s the length of the arc AP. Suppose the curve to revolve round the axis of x, and let S denote the area of the surface formed by the revolution of AP; then (*Diff. Cal.* Art. 315)

$$\frac{dS}{ds} = 2\pi y;$$

therefore $$S = \int 2\pi y \, ds \quad \ldots\ldots\ldots\ldots\ldots (1);$$

thus $$S = \int 2\pi y \frac{ds}{dx} dx \quad \ldots\ldots\ldots\ldots (2),$$

and $$S = \int 2\pi y \frac{ds}{dy} dy \quad \ldots\ldots\ldots\ldots (3).$$

Of these three forms we can choose in any particular example that which is most convenient. If y can be easily expressed in terms of s we may use (1); if $\frac{ds}{dy}$ can be easily expressed in terms of y we may use (3); in some cases it may be more convenient to express y and $\frac{ds}{dx}$ in terms of x and use (2).

In each case the area of the surface generated by the arc of the curve which lies between assigned points will be found by integrating between appropriate limits.

162. *Application to the Cylinder.*

Suppose a straight line parallel to the axis of x to revolve round the axis of x, thus generating a right circular cylinder: let a be the distance of the revolving line from the axis of x;

then $$y = a, \quad \text{and} \quad \frac{ds}{dx} = 1;$$

thus by equation (2) of Art. 161,

$$S = 2\pi \int a\, dx = 2\pi ax + C.$$

Suppose the abscissæ of the extreme points of the portion of the line which revolves to be x_1 and x_2; then the surface generated

$$= 2\pi a \int_{x_1}^{x_2} dx = 2\pi a\,(x_2 - x_1).$$

163. *Application to the Cone.*

Let a straight line which passes through the origin and is inclined to the axis of x at an angle α revolve round the axis of x, and thus generate a conical surface. Then

$$y = x \tan \alpha, \quad \text{and} \quad \frac{ds}{dx} = \sec \alpha;$$

thus by equation (2) of Art. 161,

$$S = 2\pi \int \tan \alpha \sec \alpha\, x\, dx = \pi \tan \alpha \sec \alpha\, x^2 + C.$$

Hence the surface of the frustum of a cone cut off by planes perpendicular to its axis at distances x_1, x_2 respectively from the vertex is

$$\pi \tan \alpha \sec \alpha\,(x_2^{\,2} - x_1^{\,2}).$$

Suppose $x_1 = 0$, and let r be the radius of the section made by the plane at the distance x_2, then $r = x_2 \tan \alpha$, and the area is

$$\pi \operatorname{cosec} \alpha \, r^2.$$

164. *Application to the Sphere.*

Let the circle given by the equation $y^2 = a^2 - x^2$ revolve round the axis of x; here

$$\frac{dy}{dx} = -\frac{x}{y},$$

and
$$\frac{ds}{dx} = \sqrt{\left\{1 + \left(\frac{dy}{dx}\right)^2\right\}} = \sqrt{\left(1 + \frac{x^2}{y^2}\right)} = \frac{a}{y}.$$

Hence by equation (2) of Art. 161,

$$S = 2\pi \int y \frac{a}{y} \, dx = 2\pi a \int dx = 2\pi a x + C.$$

Thus the surface included between the planes determined by

$$x = x_1 \text{ and } x = x_2 \text{ is } 2\pi a (x_2 - x_1).$$

Hence the area of a zone of a sphere depends only on the *height* of the zone and the *radius* of the sphere, and is equal to the area which the planes that bound it would cut off from a cylinder having its axis perpendicular to the planes and circumscribing the sphere; and thus the surface of the whole sphere is $4\pi a^2$. These results are very important.

165. *Application to the Prolate Spheroid.*

Let the ellipse given by $a^2y^2 + b^2x^2 = a^2b^2$ revolve round the axis of x which is supposed to coincide with the major axis of the ellipse; here

$$\frac{dy}{dx} = -\frac{b^2x}{a^2y},$$

and
$$\frac{ds}{dx} = \sqrt{\left(1 + \frac{b^4x^2}{a^4y^2}\right)} = \frac{b\sqrt{(a^2 - e^2x^2)}}{ay}.$$

Hence by equation (2) of Art. 161,

$$S = \frac{2\pi b}{a} \int \sqrt{(a^2 - e^2x^2)} \, dx = \frac{2\pi b e}{a} \int \sqrt{\left(\frac{a^2}{e^2} - x^2\right)} \, dx$$

$$= \frac{\pi b e}{a} \left\{ x \sqrt{\left(\frac{a^2}{e^2} - x^2\right)} + \frac{a^2}{e^2} \sin^{-1} \frac{ex}{a} \right\}.$$

The surface generated by the revolution of a quadrant of the ellipse will be obtained by taking 0 and a as the limits of x in the integration. This gives

$$\pi ab\left\{\sqrt{(1-e^2)}+\frac{\sin^{-1}e}{e}\right\}.$$

166. Suppose one curve to have for its equation $y=\phi(x)$, and another curve to have for its equation $y=\psi(x)$, and let both curves revolve round the axis of x. Let s_1 and s_2 denote the lengths of arcs measured from fixed points in the two curves up to the point whose abscissa is x. Let S denote the *sum* of the areas of both surfaces intercepted between two planes perpendicular to the axis of x at the distances x_1 and x_2 respectively from the origin. Then, by Art. 161,

$$S=2\pi\int_{x_1}^{x_2}\left\{\phi(x)\frac{ds_1}{dx}+\psi(x)\frac{ds_2}{dx}\right\}dx.$$

Suppose, for example, that there is a curve which is bisected by the line $y=a$, so that we may put $y=a+\chi(x)$ for the upper branch and $y=a-\chi(x)$ for the lower branch. Hence

$$\frac{ds_1}{dx}=\frac{ds_2}{dx},$$

and
$$S=4\pi a\int_{x_1}^{x_2}\frac{ds_1}{dx}\,dx=4\pi a\int ds_1,$$

the limits for s_1 being taken so as to correspond with the assigned limits of x.

Hence, if there be any complete curve which is bisected by a straight line and made to revolve round an axis which is parallel to this line at a distance a from it and which does not cut the curve, the area of the whole surface generated is equal to the length of the curve multiplied by $2\pi a$.

167. For example, take the circle given by the equation

$$(x-h)^2+(y-k)^2-c^2=0.$$

Here the area of the whole surface generated by the revolution of the circle round the axis of x will be $2\pi k\times 2\pi c$.

There is no difficulty in this example in obtaining separately the two portions of the surface. For the part above the line $y = k$, we have $2\pi\int y\,ds$, that is,

$$2\pi\int[k + \sqrt{\{c^2 - (x-h)^2\}}]\,ds,$$

that is, $$2\pi\int k\,ds + 2\pi\int\sqrt{\{c^2 - (x-h)^2\}}\,ds.$$

The former of these integrals is $2\pi ks$; the latter is equal to

$$2\pi\int\sqrt{\{c^2 - (x-h)^2\}}\,\frac{ds}{dx}\,dx,$$

which will reduce to $2\pi\int c\,dx$, that is, $2\pi cx$. Hence the surface required is found by taking the expression $2\pi ks + 2\pi cx$ between proper limits.

Area of Surfaces of Revolution. Polar Formulæ.

168. It may be sometimes convenient to use polar coordinates; thus from Art. 161 we deduce

$$S = \int 2\pi y\,ds = \int 2\pi y\,\frac{ds}{d\theta}\,d\theta = \int 2\pi r\sin\theta\,\frac{ds}{d\theta}\,d\theta,$$

where $$\frac{ds}{d\theta} = \sqrt{\left\{r^2 + \left(\frac{dr}{d\theta}\right)^2\right\}}.$$

169. *Application to the Cardioide.*

Here $r = a(1 + \cos\theta)$; thus

$$\frac{ds}{d\theta} = a\sqrt{\{(1+\cos\theta)^2 + \sin^2\theta\}} = a\sqrt{(2 + 2\cos\theta)} = 2a\cos\frac{\theta}{2};$$

therefore

$$S = 4\pi a^2\int(1+\cos\theta)\cos\frac{\theta}{2}\sin\theta\,d\theta = 16\pi a^2\int\cos^4\frac{\theta}{2}\sin\frac{\theta}{2}\,d\theta$$

$$= -\frac{32\pi a^2}{5}\cos^5\frac{\theta}{2} + C.$$

The surface formed by the revolution of the complete curve about the initial line will be obtained by taking 0 and π as the limits of θ in the integral. This gives $\frac{32\pi a^2}{5}$.

Any Surface. Double Integration.

170. Let x, y, z be the co-ordinates of any point p of a surface; $x+\Delta x$, $y+\Delta y$, $z+\Delta z$ the co-ordinates of an adjacent point q. Through p draw a plane parallel to that of (x, z), and a plane parallel to that of (y, z); also through q draw a plane parallel to that of (x, z) and a plane parallel to that of (y, z). These planes will intercept an element pq of the curved surface, and the projection of this element on the plane of (x, y) will be the rectangle PQ. Suppose the tangent plane to the surface at p to be inclined to the plane of (x, y) at an angle γ, then it is known from solid geometry that

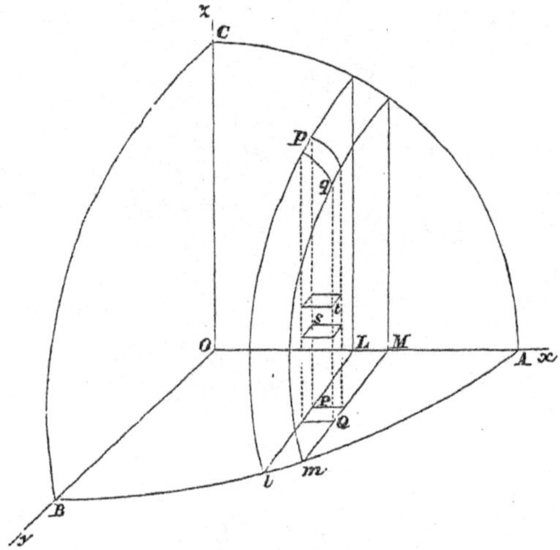

$$\sec\gamma = \sqrt{\left\{1+\left(\frac{dz}{dx}\right)^2+\left(\frac{dz}{dy}\right)^2\right\}},$$

where $\frac{dz}{dx}$ and $\frac{dz}{dy}$ must be found from the known equation to the surface. Now the area of PQ is $\Delta x\,\Delta y$, hence by solid geometry the area of the element of the tangent plane at p of which PQ is the projection is $\Delta x \Delta y \sec\gamma$. We shall assume that the limit of the sum of such terms as $\Delta x \Delta y \sec\gamma$ for all values of x and y comprised between assigned limits is the area of the surface corresponding to those limits. Let then S denote this surface; thus

$$S = \iint \sqrt{\left\{1 + \left(\frac{dz}{dx}\right)^2 + \left(\frac{dz}{dy}\right)^2\right\}}\, dx\, dy,$$

the limits of the integrations being dependent upon the portion of the surface considered.

171. With respect to the point assumed in the preceding article, the reader is referred to the remarks on a similar point in *Diff. Cal.* Art. 308. He may also hereafter consult De Morgan's *Differential and Integral Calculus*, page 444, and Homersham Cox's *Integral Calculus*, page 96.

172. *Application to the Sphere.*

Let it be required to find the area of the eighth part of the surface of the sphere given by the equation

$$x^2 + y^2 + z^2 = a^2.$$

Here
$$\frac{dz}{dx} = -\frac{x}{z}, \qquad \frac{dz}{dy} = -\frac{y}{z};$$

thus
$$S = \iint \sqrt{\left(1 + \frac{x^2}{z^2} + \frac{y^2}{z^2}\right)}\, dx\, dy = \iint \frac{a\, dx\, dy}{\sqrt{(a^2 - x^2 - y^2)}}.$$

Now in the figure we suppose $OL = x$; put y_1 for Ll, then $y_1 = \sqrt{(a^2 - x^2)}$, for the value of y_1 is obtained from the equation to the surface by supposing $z = 0$. If we integrate with respect to y between the limits 0 and y_1, we sum up all the elements comprised in a strip of which $LMml$ is the projection on the plane of (x, y). Now

$$\int_0^{y_1} \frac{dy}{\sqrt{(a^2 - x^2 - y^2)}} = \int_0^{y_1} \frac{dy}{\sqrt{(y_1^2 - y^2)}} = \frac{\pi}{2};$$

thus
$$S=\frac{\pi a}{2}\int dx.$$

If we integrate with respect to x from 0 to a, we sum up all the strips comprised in the surface of which OAB is the projection. Thus $\frac{\pi a^2}{2}$ is the required result; and therefore the whole surface of the sphere is $4\pi a^2$.

If we integrate with respect to x first, we shall have
$$S=\int_0^a\int_0^{x_1}\frac{a\,dy\,dx}{\sqrt{(a^2-x^2-y^2)}},$$
where $x_1=\sqrt{(a^2-y^2)}$.

173. As another example let it be required to find the area of that part of the surface given by the equation
$$z^2+(x\cos\alpha+y\sin\alpha)^2-a^2=0,$$
which is situated in the positive compartment of co-ordinates. This surface is a right circular cylinder, having for its axis the line determined by $z=0$, $x\cos\alpha+y\sin\alpha=0$, and a is the radius of a circular section of it. Here
$$\frac{dz}{dx}=-\frac{\cos\alpha\,(x\cos\alpha+y\sin\alpha)}{z},$$
$$\frac{dz}{dy}=-\frac{\sin\alpha\,(x\cos\alpha+y\sin\alpha)}{z},$$
thus
$$S=\iint\frac{a\,dx\,dy}{z}=\iint\frac{a\,dx\,dy}{\sqrt{\{a^2-(x\cos\alpha+y\sin\alpha)^2\}}}.$$

The co-ordinate plane of (x, y) cuts the surface in the straight lines $a=\pm(x\cos\alpha+y\sin\alpha)$, and if the upper sign be taken, we have a line lying in the positive quadrant of the plane of (x, y).

To obtain the value of S we integrate first with respect to y between the limits $y=0$ and $y=(a-x\cos\alpha)\operatorname{cosec}\alpha$; now
$$\int\frac{dy}{\sqrt{\{a^2-(x\cos\alpha+y\sin\alpha)^2\}}}=\frac{1}{\sin\alpha}\sin^{-1}\frac{x\cos\alpha+y\sin\alpha}{a};$$

take this between the assigned limits, and we obtain

$$\frac{1}{\sin \alpha}\left(\frac{\pi}{2} - \sin^{-1}\frac{x \cos \alpha}{a}\right);$$

therefore $$S = \frac{a}{\sin \alpha}\int\left\{\frac{\pi}{2} - \sin^{-1}\frac{x \cos \alpha}{a}\right\} dx,$$

and the limits of the integration are 0 and $\frac{a}{\cos \alpha}$. Hence we shall find

$$S = \frac{a^2}{\sin \alpha \cos \alpha}.$$

174. Instead of taking the element of the tangent plane at any point of a surface, so that its projection shall be the rectangle $\Delta x \, \Delta y$, it may be in some cases more convenient to take it so that its projection shall be the polar element $r\Delta\theta \, \Delta r$. Thus we shall have

$$S = \iint \sec \gamma \, r d\theta \, dr.$$

For example, suppose we require the area of the surface $xy = az$, which is cut off by the surface $x^2 + y^2 = c^2$; here

$$\sec \gamma = \sqrt{\left(1 + \frac{x^2}{a^2} + \frac{y^2}{a^2}\right)} = \frac{\sqrt{(a^2 + r^2)}}{a} \text{ since } x^2 + y^2 = r^2.$$

Thus $$S = \int_0^{2\pi}\int_0^c \frac{\sqrt{(a^2 + r^2)}}{a} r d\theta \, dr = \frac{2\pi}{3a}\{(c^2 + a^2)^{\frac{3}{2}} - a^3\}.$$

175. Suppose $x = r \sin \theta \cos \phi$, $y = r \sin \theta \sin \phi$, $z = r \cos \theta$, so that r, θ, ϕ are the usual polar co-ordinates of a point in space; then we shall shew hereafter that the equation

$$S = \iint \sqrt{\left\{1 + \left(\frac{dz}{dx}\right)^2 + \left(\frac{dz}{dy}\right)^2\right\}} dx \, dy$$

may be transformed into

$$S = \iint \sqrt{\left\{r^2 \sin^2 \theta + \left(\frac{dr}{d\theta}\right)^2 \sin^2 \theta + \left(\frac{dr}{d\phi}\right)^2\right\}} r d\theta \, d\phi.$$

An independent geometrical proof will be found in the *Cambridge and Dublin Mathematical Journal*, Vol. IX., and also in Carmichael's *Treatise on the Calculus of Operations.* It will be remembered that in this formula $r = \sqrt{(x^2 + y^2 + z^2)}$, while in Art. 174 we denote $\sqrt{(x^2 + y^2)}$ by r.

Approximate Values of Integrals.

176. Suppose y a function of x, and that we require $\int_a^c y\,dx$. If the *indefinite* integral $\int y\,dx$ is known we can at once ascertain the required definite integral. If the indefinite integral is unknown, we may still determine approximately the value of the definite integral. This process of approximation is best illustrated by supposing y to be an ordinate of a curve so that $\int_a^c y\,dx$ represents a certain area. Divide $c - a$ into n parts each equal to h and draw $n - 1$ ordinates at equal distances between the initial and final ordinates; then the ordinates may be denoted by $y_1, y_2, \ldots\ldots y_n, y_{n+1}$. Hence we may take

$$h(y_1 + y_2 + \ldots\ldots + y_n)$$

as an approximate value of the required area. Or we may take

$$h(y_2 + y_3 \ldots\ldots + y_{n+1})$$

as an approximate value.

We may obtain another approximation thus; suppose the extremities of the rth and $\overline{r+1}|^{\text{th}}$ ordinates joined; thus we have a trapezium, the area of which is $(y_r + y_{r+1})\frac{h}{2}$. The sum of all such trapeziums gives as an approximate value of the area

$$h\left\{\frac{y_1}{2} + y_2 + y_3 \ldots\ldots + y_n + \frac{y_{n+1}}{2}\right\}.$$

This result is in fact half the sum of the two former results. It is obvious we may make the approximation as close as we please by sufficiently increasing n.

177. The following is another method of approximation. Let a parabola be drawn having its axis parallel to that of y; let y_1, y_2, y_3 represent three equidistant ordinates, h the distance between y_1 and y_2, and therefore also between y_2 and y_3. Then it may be proved that the area contained between the parabola, the axis of x, and the two extreme ordinates is

$$\frac{h}{3}(y_1 + 4y_2 + y_3).$$

This will be easily shewn by a figure, as the area consists of a trapezium and a parabolic segment, and the area of the latter is known by Art. 143.

Let us now suppose that n is even, so that the area we have to estimate is divided into an even number of pieces. Then assume that the area of the first two pieces is

$$\frac{h}{3}(y_1 + 4y_2 + y_3),$$

that the area of the third and fourth pieces is

$$\frac{h}{3}(y_3 + 4y_4 + y_5),$$

and so on. Thus we shall have finally as an approximate result

$$\frac{h}{3}\{y_1 + 2(y_3 + y_5 + \ldots\ldots y_{n-1}) + y_{n+1} + 4(y_2 + y_4 \ldots\ldots + y_n)\}.$$

Hence we have the following rule: add together the first ordinate, the last ordinate, twice the sum of all the other odd ordinates, and four times the sum of all the even ordinates; then multiply the result by one-third the common distance of the ordinates. This rule is called *Simpson's Rule.*

EXAMPLES.

1. If A denote the area contained between the catenary, the axis of x, the axis of y, and an ordinate at the extremity of the arc s, shew that $A = cs$. The arc s begins at the lowest point of the curve.

2. The whole area of the curve

$$\left(\frac{x}{a}\right)^{\frac{2}{3}} + \left(\frac{y}{b}\right)^{\frac{2}{3}} = 1$$

is $\frac{3}{8}\pi ab$. (The integration may be effected by assuming $x = a \cos^3\phi$.)

3. The area of the curve $y(x^2 + a^2) = c^2(a - x)$ from $x = 0$ to $x = a$ is $c^2\left(\frac{\pi}{4} - \frac{1}{2}\log 2\right)$.

4. The area of the curve $y^2x = 4a^2(2a - x)$ from $x = 0$ to $x = 2a$ is $4\pi a^2$.

5. Find the whole area between the curve $y^2(x^2 + a^2) = a^2x^2$ and its asymptotes. *Result.* $4a^2$.

6. Find the area of the loop of the curve $y^2 = \frac{x^2(a + x)}{a - x}$.

Result. $2a^2\left(1 - \frac{\pi}{4}\right)$.

7. Find the area bounded by the curve $y^2 = \frac{x^2(a + x)}{a - x}$ and the asymptote $x = a$, excluding the loop.

Result. $2a^2\left(1 + \frac{\pi}{4}\right)$.

8. Find the whole area between the curve $y^2(2a - x) = x^3$ and its asymptote. *Result.* $3\pi a^2$.

9. Find the whole area of the curve $(y - x)^2 = a^2 - x^2$. *Result.* πa^2.

10. Find the area included between the curves

$$y^2 - 4ax = 0, \quad x^2 - 4ay = 0. \qquad \textit{Result.}\ \frac{16a^2}{3}.$$

11. Find the whole area of the curve $a^4y^2 + b^2x^4 = a^2b^2x^2$. *Result.* $\frac{4}{3}ab$.

12. Find the area of a loop of the curve $a^2y^4 = x^4(a^2 - x^2)$.

Result. $\dfrac{4a^2}{5}$.

13. The area between the tractory, the axis of y, and the asymptote is $\dfrac{\pi c^2}{4}$. (See Art. 100.)

14. Find the area of a loop of the curve

$$y^2(a^2 + x^2) = x^2(a^2 - x^2). \quad \textit{Result.}\ \frac{a^2}{2}(\pi - 2).$$

15. Find the area of the loop of the curve

$$16a^4y^2 = b^2x^2(a^2 - 2ax). \qquad \textit{Result.}\ \frac{ab}{30}.$$

16. Find the area of the loop of the curve

$$2y^2(a^2 + x^2) = (a^2 - x^2)^2.$$

Result. $a^2\{3\sqrt{2}\log(1 + \sqrt{2}) - 2\}$.

17. Find the whole area of the curve

$$2y^2(a^2 + x^2) - 4ay(a^2 - x^2) + (a^2 - x^2)^2 = 0.$$

Result. $a^2\pi\left\{4 - \dfrac{5\sqrt{2}}{2}\right\}$.

18. Find the area of the curve

$$y = c\sin\frac{x}{a}\,.\log\sin\frac{x}{a}$$

from $x = 0$ to $x = a\pi$. *Result.* $2ac(1 - \log 2)$.

19. Find the area of the curve $\dfrac{y}{c} = \left(\dfrac{x}{a}\right)^n$ between $x = \alpha$ and $x = \beta$, and *from the result* deduce the area of the hyperbola $xy = a^2$ between the same limits.

20. Find the area of the ellipse whose equation is

$$ax^2 + 2bxy + cy^2 = 1. \qquad \textit{Result.}\ \frac{\pi}{\sqrt{(ac - b^2)}}.$$

21. Find the area of a loop of the curve $r^2 = a^2 \cos 2\theta$.

Result. $\dfrac{a^2}{2}$.

22. Find the area contained by all the loops of the curve

$$r = a \sin n\theta.$$

Result. $\dfrac{\pi a^2}{4}$ or $\dfrac{\pi a^2}{2}$ according as n is odd or even.

23. Find the area between the curves $r = a \cos n\theta$ and $r = a$.

24. Find the area of a loop of the curve $r^2 \cos \theta = a^2 \sin 3\theta$.

Result. $\dfrac{3a^2}{4} - \dfrac{a^2}{2} \log 2$.

25. Find the whole area of the curve $r = a(\cos 2\theta + \sin 2\theta)$.

Result. πa^2.

26. Find the area of a loop of the curve $(x^2 + y^2)^3 = 4a^2x^2y^2$.

Result. $\dfrac{\pi a^2}{8}$.

27. Find the whole area of the curve

$(x^2 + y^2)^2 = 4a^2x^2 + 4b^2y^2$. *Result.* $2\pi(a^2 + b^2)$.

28. Find the whole area of the curve

$\dfrac{x^2}{a^4} + \dfrac{y^2}{b^4} = \dfrac{1}{c^2}\left(\dfrac{x^2}{a^2} + \dfrac{y^2}{b^2}\right)^2$. *Result.* $\dfrac{\pi c^2}{2ab}(a^2 + b^2)$.

29. Find the area of the loop of the curve

$y^3 - 3axy + x^3 = 0$. *Result.* $\dfrac{3a^2}{2}$.

30. Find the area of the loop of the curve

$r \cos \theta = a \cos 2\theta$. *Result.* $\left(2 - \dfrac{\pi}{2}\right) a^2$.

31. Find the area of the curve

$$r = \frac{a^2}{\sqrt{(a^2 - b^2 \cos^2 \theta)}} + b \cos \theta,$$

a being greater than b. *Result.* $\dfrac{\pi a^3}{\sqrt{(a^2 - b^2)}} + \dfrac{\pi b^2}{2}$.

32. In a logarithmic spiral find the area between the curve and two radii vectores drawn from the pole.

33. Find the area between the conchoid $r = a + b \operatorname{cosec} \theta$ and two radii vectores drawn from the pole.

34. In an ellipse find the area between the curve and two radii vectores drawn from the centre.

35. In a parabola find the area between the curve and two radii vectores drawn from the vertex.

36. Find the area included between the curve

$$r = a (\sec \theta + \tan \theta)$$

and its asymptote $r \cos \theta = 2a$. *Result.* $\left(\dfrac{\pi}{2} + 2\right) a^2$.

37. The whole area of the curve $r = a (2 \cos \theta + 1)$ is $a^2 \left(2\pi + \dfrac{3\sqrt{3}}{2}\right)$, and the area of the inner loop is $a^2 \left(\pi - \dfrac{3\sqrt{3}}{2}\right)$.

38. Find the whole area of the curve $r = a \cos \theta + b$, where a is greater than b. Also find the area of the inner loop.

39. If x and y be the co-ordinates of an equilateral hyperbola $x^2 - y^2 = a^2$, shew that

$$x = \frac{a}{2} \left(e^{\frac{2u}{a^2}} + e^{-\frac{2u}{a^2}}\right), \quad y = \frac{a}{2} \left(e^{\frac{2u}{a^2}} - e^{-\frac{2u}{a^2}}\right),$$

where u is the area intercepted between the curve, the central radius vector, and the axis.

40. Find the whole area of the curve which is the locus of the intersection of two normals to an ellipse at right angles. *Result.* $\pi(a-b)^2$.

It may be shewn that the equation to the curve is

$$r^2(a^2+b^2)(a^2\sin^2\theta+b^2\cos^2\theta)^2$$
$$=(a^2-b^2)^2(a^2\sin^2\theta-b^2\cos^2\theta)^2.$$

41. Find the area included within any arc traced by the extremity of the radius vector of a spiral in a complete revolution, and the straight line joining the extremities of the arc. If, for example, the equation to the spiral be $r=a\left(\frac{\theta}{2\pi}\right)^n$, prove that the area corresponding to any value of θ greater than 2π is

$$\frac{\pi a^2}{2n+1}\left\{\left(\frac{\theta}{2\pi}\right)^{2n+1}-\left(\frac{\theta}{2\pi}-1\right)^{2n+1}\right\}.$$

42. Find the area contained between a parabola, its evolute, and two radii of curvature of the parabola. (Art. 159.)

43. Find the area contained between a cycloid, its evolute, and two radii of curvature of the cycloid.

44. Find the area of the surface generated by the revolution round the axis of x of the curve $xy=k^2$.

45. Also of the curve $y=ae^{\frac{x}{c}}$.

46. Also of the catenary $y=\frac{c}{2}(e^{\frac{x}{c}}+e^{-\frac{x}{c}})$.

47. Shew that the whole surface of an oblate spheroid is

$$2\pi a^2\left\{1+\frac{1-e^2}{2e}\log\frac{1+e}{1-e}\right\}.$$

48. A cycloid revolves round the tangent at the vertex; shew that the whole surface generated is $\frac{32}{3}\pi a^2$.

49. A cycloid revolves round its base; shew that the whole surface generated is $\frac{64}{3}\pi a^2$.

50. A cycloid revolves round its axis; shew that the whole surface generated is $8\pi a^2(\pi - \frac{4}{3})$.

51. The whole surface generated by the revolution of the tractory round the axis of x is $4\pi c^2$.

52. A sphere is pierced perpendicularly to the plane of one of its great circles by two right cylinders, of which the diameters are equal to the radius of the sphere and the axes pass through the middle points of two radii that compose a diameter of this great circle. Find the surface of that portion of the sphere not included within the cylinders.

Result. Twice the square of the diameter of the sphere.

53. Find the surface generated by the portion of the curve $y = a + a\log\frac{x}{a}$ between the limits $x = a$ and $x = ae$.

$$\textit{Result.}\quad 4\pi a^2\left\{1 + \sqrt{(1+e^2)} - \sqrt{2} + \log\frac{1+\sqrt{2}}{1+\sqrt{(1+e^2)}}\right\}.$$

54. Find $\int\frac{dS}{p}$, where dS represents an element of surface, and p the perpendicular from the origin upon the tangent plane of the element, the integral being extended over the whole of the ellipsoid $\frac{x^2}{a^2} + \frac{y^2}{b^2} + \frac{z^2}{c^2} = 1$.

$$\textit{Result.}\quad \frac{4\pi}{3abc}(a^2b^2 + b^2c^2 + c^2a^2).$$

CHAPTER VIII.

VOLUMES OF SOLIDS.

Formulæ involving Single Integration. Solid of Revolution.

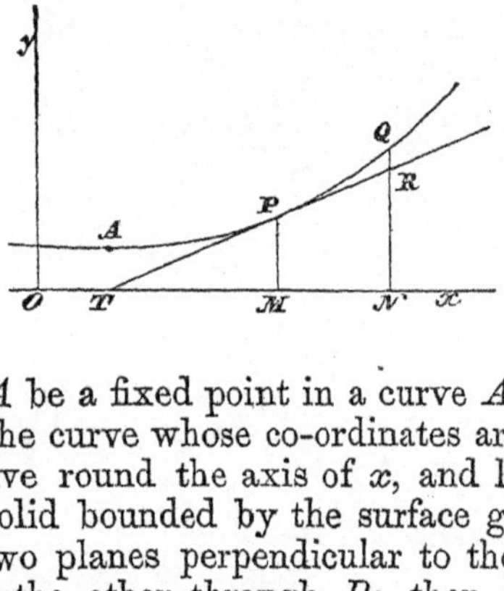

178. LET A be a fixed point in a curve APQ, and P any other point on the curve whose co-ordinates are x and y. Let the curve revolve round the axis of x, and let V denote the volume of the solid bounded by the surface generated by the curve and by two planes perpendicular to the axis of x, one through A and the other through P; then (*Diff. Cal.* Art. 314)

$$\frac{dV}{dx} = \pi y^2,$$

therefore

$$V = \int \pi y^2 dx.$$

From the equation to the curve y is a known function of x; suppose $\psi(x)$ to be the integral of πy^2; then

$$V = \psi(x) + C.$$

Let V_1 denote the volume when the point P has x_1 for its abscissa, and V_2 the volume when the point P has x_2 for its abscissa; thus

$$V_1 = \psi(x_1) + C,$$

$$V_2 = \psi(x_2) + C,$$

therefore $$V_2 - V_1 = \psi(x_2) - \psi(x_1) = \pi \int_{x_1}^{x_2} y^2 dx.$$

179. *Application to the Right Circular Cone.*

Let a straight line pass through the origin and make an angle α with the axis of x; then this straight line will generate a right circular cone by revolving round the axis of x. Here $y = x \tan \alpha$; thus

$$V = \int \pi \tan^2 \alpha \, x^2 dx = \frac{\pi \tan^2 \alpha}{3} x^3 + C,$$

$$V_2 - V_1 = \frac{\pi \tan^2 \alpha}{3} (x_2^3 - x_1^3).$$

Suppose $x_1 = 0$, and let $r = x_2 \tan \alpha$; thus the volume becomes $\frac{\pi \tan^2 \alpha \, x_2^3}{3}$, that is, $\frac{\pi r^2 x_2}{3}$. Hence the volume of a right circular cone is one-third the product of the area of the base into the altitude.

180. *Application to the Sphere.*

Here taking the origin at the centre of the sphere we have $y^2 = a^2 - x^2$; thus

$$\int \pi y^2 dx = \pi \left(a^2 x - \frac{x^3}{3}\right) + C.$$

The volume of a hemisphere $= \int_0^a \pi y^2 dx = \frac{2\pi a^3}{3}$.

181. *Application to the Paraboloid.*

Here the generating curve is the parabola, so that

$$y^2 = 4ax.$$

Thus $$V_2 - V_1 = \pi \int_{x_1}^{x_2} 4ax\,dx = 2a\pi\,(x_2^2 - x_1^2).$$

Suppose $x_1 = 0$, then the volume becomes $2a\pi x_2^2$, that is $\frac{1}{2}\pi y_2^2 x_2$, where $y_2^2 = 4ax_2$; thus the volume is half that of a cylinder which has the same height, namely x_2, and the same base, namely a circle of which y_2 is the radius.

182. *Application to the Solid formed by a Cycloid.*

Let a cycloid revolve round its axis; here (*Diff. Cal.* Art. 358)

$$y = \sqrt{(2ax - x^2)} + a\,\text{vers}^{-1}\frac{x}{a}.$$

The integration is best effected by putting for x and y their values in terms of θ (*Diff. Cal.* Art. 358). Thus

$$\pi\int y^2 dx = \pi a^3 \int (\theta + \sin\theta)^2 \sin\theta\,d\theta.$$

To obtain the volume generated by a semi-cycloid the limits for x would be 0 and $2a$; thus the corresponding limits for θ are 0 and π.

Now $$\int \theta^2 \sin\theta\,d\theta = -\theta^2\cos\theta + 2\int\theta\cos\theta\,d\theta$$

$$= -\theta^2\cos\theta + 2\theta\sin\theta + 2\cos\theta,$$

therefore $$\int_0^{\pi} \theta^2\sin\theta\,d\theta = \pi^2 - 4;$$

$$2\int\theta\sin^2\theta\,d\theta = \int\theta\,(1 - \cos 2\theta)\,d\theta = \frac{\theta^2}{2} - \frac{\theta\sin 2\theta}{2} - \frac{\cos 2\theta}{4},$$

therefore $$2\int_0^{\pi}\theta\sin^2\theta\,d\theta = \frac{\pi^2}{2}.$$

And $$\int_0^{\pi}\sin^3\theta\,d\theta = 2\int_0^{\frac{\pi}{2}}\sin^3\theta\,d\theta = 2\,.\,\frac{2}{3}. \quad \text{(Art. 35.)}$$

Thus the required volume

$$= \pi a^3\left\{\pi^2 - 4 + \frac{\pi^2}{2} + \frac{4}{3}\right\} = \pi a^3\left(\frac{3\pi^2}{2} - \frac{8}{3}\right).$$

183. This formula for the volume of a solid of revolution, $V = \int \pi y^2 dx$, like others which we have noticed, is one, the truth of which is obvious, as soon as the notation of the Integral Calculus is understood. In the figure to Art. 128, if PM be y and MN be denoted by Δx, then $\pi y^2 \Delta x$ is the volume of the solid generated by the revolution of $MNpP$ about the axis of x. Thus $\Sigma \pi y^2 \Delta x$ will differ from the volume generated by the revolution of $ADEB$ by the sum of such volumes as are generated by PpQ, and the latter sum will vanish in the limit. Thus the volume generated by the revolution of $ADEB$ is equal to the limit of $\Sigma \pi y^2 \Delta x$, that is, to $\int \pi y^2 \, dx$.

184. Similarly, if V denote the volume bounded by the surface formed by a curve which revolves round the axis of y, and by planes perpendicular to the axis of y, we shall have

$$V = \int \pi x^2 \, dy.$$

And, as in Art. 178, we shall have

$$V_2 - V_1 = \int_{y_1}^{y_2} \pi x^2 \, dy.$$

185. Suppose two curves to revolve round the axis of x, and thus to generate two surfaces, and that we require the *difference* of two volumes, one bounded by the first surface and by planes perpendicular to the axis of x, and the other bounded by the second surface and by the planes already assigned. Let $y = \phi(x)$ be the equation to the first curve, and $y = \psi(x)$ that to the second. Then if V denote the required difference, we have

$$V = \int \pi \{\phi(x)\}^2 \, dx - \int \pi \{\psi(x)\}^2 \, dx$$

$$= \pi \int [\{\phi(x)\}^2 - \{\psi(x)\}^2] \, dx.$$

If the planes which bound the required volume are determined by $x = x_1$ and $x = x_2$, we must integrate between the limits x_1 and x_2 for x.

186. Suppose, for example, that a closed curve is such that the line $y = a$ bisects every ordinate parallel to the axis of y; then we have $\phi(x) = a + \chi(x)$ and $\psi(x) = a - \chi(x)$, where $\chi(x)$ denotes some function of x. Thus

$$\{\phi(x)\}^2 - \{\psi(x)\}^2 = 4a\chi(x),$$

and $$V = \pi \int 4a\chi(x)\, dx.$$

Suppose the abscissæ of the extreme points of the curve are x_2 and x_1, then the volume generated by the revolution of the closed curve round the axis of x is $4a\pi \int_{x_1}^{x_2} \chi(x)\, dx$. And $2\int_{x_1}^{x_2} \chi(x)\, dx$ is the area of the closed curve, so that the volume is equal to the product of $2a\pi$ into the area. This demonstration supposes that the generating curve lies entirely on one side of the axis of x.

If the generating curve be the circle given by

$$(x - h)^2 + (y - k)^2 = c^2,$$

we have πc^2 for its area, and therefore $2kc^2\pi^2$ for the volume generated by the revolution of it round the axis of x.

187. In a similar way if the curves $x = \phi(y)$, $x = \psi(y)$, revolve round the axis of y we obtain for the volume bounded by these surfaces and by planes perpendicular to the axis of y

$$V = \pi \int [\{\phi(y)\}^2 - \{\psi(y)\}^2]\, dy.$$

188. The method given in Art. 178 for finding the volume of a *solid of revolution* may be adapted to *any* solid. The method may be described thus: conceive the solid cut up into thin slices by a series of parallel planes, estimate approximately the volume of each slice and add these volumes; the limit of this sum when each slice becomes indefinitely thin is

the volume of the solid required. Suppose that a solid is cut up into slices by planes perpendicular to the axis of x; let $\phi(x)$ be the area of a section of the solid made by a plane which is at a distance x from the origin, and let $x+\Delta x$ be the distance of the next plane from the origin; thus these two planes intercept a slice of which the thickness is Δx, and of which the volume may be represented by $\phi(x)\,\Delta x$. The volume of the solid will therefore be the limit of $\Sigma\phi(x)\,\Delta x$, that is, it will be $\int \phi(x)\,dx$; the limits of the integration will depend upon the particular solid or portion of a solid under consideration.

189. *Application to an Ellipsoid.*

The equation to the ellipsoid is

$$\frac{x^2}{a^2}+\frac{y^2}{b^2}+\frac{z^2}{c^2}=1;$$

if a section be made by a plane perpendicular to the axis of x at a distance x from the origin, the boundary of the section is an ellipse, of which the semiaxes are $b\sqrt{\left(1-\frac{x^2}{a^2}\right)}$ and $c\sqrt{\left(1-\frac{x^2}{a^2}\right)}$; hence the area of this ellipse is $\pi bc\left(1-\frac{x^2}{a^2}\right)$; this is therefore the value of $\phi(x)$. Hence the volume of the ellipsoid

$$=\int_{-a}^{a}\pi bc\left(1-\frac{x^2}{a^2}\right)dx=\frac{4\pi abc}{3}.$$

190. *Application to a Pyramid.*

Let there be a pyramid, the base of which is any rectilinear figure; let A be the area of the base and h the height. Take the origin of co-ordinates at the vertex of the pyramid, and the axis of x perpendicular to the base of the pyramid, then the volume of the pyramid

$$=\int_0^h \phi(x)\,dx.$$

Now the section of the pyramid made by any plane parallel to the base is a rectilinear figure similar to the base, and the areas of similar figures are as the squares of their homologous sides; and x and h are proportional to homologous sides; hence we infer that

$$\phi(x) = \frac{x^2}{h^2}A.$$

Thus the volume of the pyramid

$$= \frac{A}{h^2}\int_0^h x^2 dx = \frac{Ah}{3}.$$

This investigation also holds for a cone, the base of which is any closed curve.

191. As an example we will find the volume lying between an hyperboloid of one sheet, its asymptotic cone and two planes perpendicular to their common axis.

Let the equation to the hyperboloid be

$$\frac{x^2}{a^2} - \frac{y^2}{b^2} - \frac{z^2}{c^2} + 1 = 0,$$

and that to the cone

$$\frac{x^2}{a^2} - \frac{y^2}{b^2} - \frac{z^2}{c^2} = 0.$$

If a section of the former surface be made by a plane perpendicular to the axis of x and at a distance x from the origin, the boundary is an ellipse of which the area is $\pi bc\left(\frac{x^2}{a^2} + 1\right)$; the section of the second surface made by the same plane also has an ellipse for its boundary, and its area is $\frac{\pi bcx^2}{a^2}$. Therefore the difference of the areas is πbc. Hence the required volume, supposing it bounded by the planes $x = x_1$ and $x = x_2$, is

$$\int_{x_1}^{x_2} \pi bc\, dx, \text{ that is, } \pi bc\,(x_2 - x_1).$$

192. Sometimes it may be convenient to make sections by parallel planes not perpendicular to the axis of x. If α be the inclination of the axis of x to the parallel planes, then $\phi(x)\sin\alpha\Delta x$ may be taken as the volume of a slice and the integration performed as before.

Formulæ involving Double Integration.

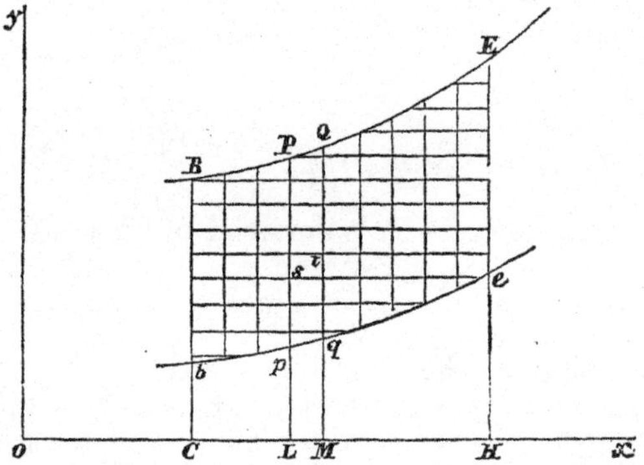

193. We will first give a formula for the volume of a solid of revolution. In the figure, let x, y be the co-ordinates of s, and $x+\Delta x$, $y+\Delta y$ those of t. Suppose the whole figure to revolve round the axis of x, then the element st will generate a ring, the volume of which will be ultimately $2\pi y\,\Delta x\,\Delta y$: this follows from the consideration that $\Delta x\,\Delta y$ is the area of st and $2\pi y$ the perimeter of the circle described by s. Hence the volume generated by the figure $BEeb$, or by any portion of it, will be the limit of the sum of such terms as $2\pi y\Delta x\Delta y$. Let V denote the required volume, then

$$V = 2\pi \iint y\,dx\,dy;$$

the limits of the integration being so taken as to include all the elements of the required volume.

194. Suppose that the volume required is that which is obtained by the revolution of all the figure $BEeb$; let $y=\phi(x)$

be the equation to the upper curve, $y=\psi(x)$ that to the lower curve, and let $OC=x_1$, $OH=x_2$. We should then integrate first with respect to y between the limits $y=\psi(x)$ and $y=\phi(x)$; we thus sum up all the elements like $2\pi y\,\Delta x\,\Delta y$ which are contained in the solid formed by the revolution of the strip $PQqp$; then we integrate with respect to x between the limits x_1 and x_2. Thus to express the operation symbolically

$$V=2\pi\int_{x_1}^{x_2}\int_{\psi(x)}^{\phi(x)} y\,dx\,dy$$

$$=\pi\int_{x_1}^{x_2}[\{\phi(x)\}^2-\{\psi(x)\}^2]\,dx.$$

The second expression is obtained by effecting the integration with respect to y between the assigned limits, and it coincides with that already obtained in Art. 185.

195. Thus in the preceding article we divide the solid into elementary rings, of which $2\pi y\,\Delta x\,\Delta y$ is the type; in the first integration we collect a number of these rings, so as to form a figure, which is the difference of two concentric circular slices; in the second integration we collect all these figures and thus obtain the volume of the required solid.

196. Suppose the figure which revolves round the axis of x to be bounded by the curves $x=\phi(y)$ and $x=\psi(y)$, and by the straight lines $y=y_1$ and $y=y_2$; then in applying the formula for V it will be convenient to integrate first with respect to x; thus

$$V=2\pi\int_{y_1}^{y_2}\int_{\psi(y)}^{\phi(y)} y\,dy\,dx.$$

In this case in the integration with respect to x we collect all the elements like $2\pi y\,\Delta y\,\Delta x$ which have the same radius y, so that the sum of the elements is a thin cylindrical shell, of which Δy is the thickness, y is the radius, and $\phi(y)-\psi(y)$ the height. Thus

$$V=2\pi\int_{y_1}^{y_2}\{\phi(y)-\psi(y)\}\,y\,dy.$$

197. As an example of the preceding formulæ, let it be required to find the volume of the solid generated by the revolution of the area ALB round the axis of x in the figure already given in Art. 141. This volume is the excess of the hemisphere generated by the revolution of SLB over the paraboloid generated by the revolution of ASL; the result is therefore known, and we propose the example, not for the sake of the result, but for illustration of the formulæ of double integration.

Let S be the origin. Suppose the positive direction of the axis of x to the left, then the equation to AL is $y^2 = 4a(a-x)$ and that to BL is $y^2 = 4a^2 - x^2$. Let V be the required volume, then

$$V = \int_0^{2a} \int_{\frac{4a^2-y^2}{4a}}^{\sqrt{(4a^2-y^2)}} 2\pi y \, dy \, dx.$$

If we wish to integrate with respect to y first, we must, as in Art. 141, suppose the figure ALB divided into two parts; thus

$$V = \int_0^a \int_{\sqrt{(4a^2-4ax)}}^{\sqrt{(4a^2-x^2)}} 2\pi y \, dx \, dy + \int_a^{2a} \int_0^{\sqrt{(4a^2-x^2)}} 2\pi y \, dx \, dy.$$

Again, let it be required to find the volume generated by the revolution of LDC about the axis of x. Let the positive direction of the axis of x be now to the right, then the equation to LC is $y^2 = 4a(a+x)$ and that to LD is $y^2 = 4a^2 - x^2$. Let V be the required volume, then

$$V = \int_0^{2a} \int_{\sqrt{(4a^2-x^2)}}^{\sqrt{(4a^2+4ax)}} 2\pi y \, dx \, dy.$$

If we wish to integrate with respect to x first, we must, as in Art. 141, suppose the figure LDC divided into two parts; thus

$$V = \int_0^{2a} \int_{\sqrt{(4a^2-y^2)}}^{2a} 2\pi y \, dy \, dx + \int_{2a}^{2a\sqrt{3}} \int_{\frac{y^2-4a^2}{4a}}^{2a} 2\pi y \, dy \, dx.$$

198. Similarly, if a solid is formed by the revolution of a curve round the axis of y, we have

$$V = \iint 2\pi x \, dy \, dx.$$

199. We now proceed to consider any solid.

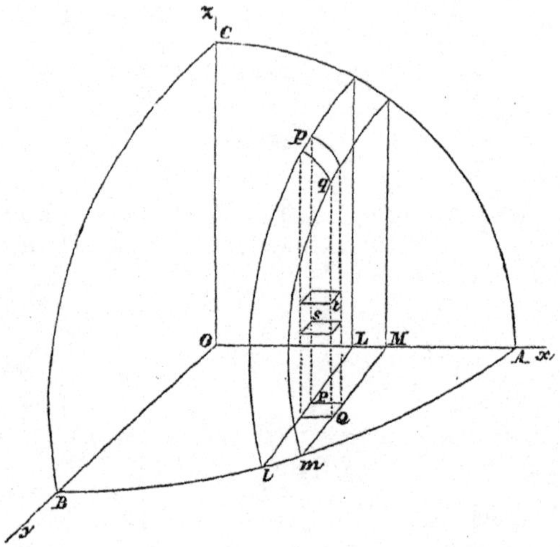

Let x, y, z be the co-ordinates of any point p of a surface, $x + \Delta x$, $y + \Delta y$, $z + \Delta z$ the co-ordinates of an adjacent point q. Through p draw planes parallel to the co-ordinate planes of (x, z) and (y, z); through q also draw planes parallel to the same co-ordinate planes. These four planes will include between them a column, of which PQ is the base and Pp the height. The volume of this column will be ultimately $z \Delta x \Delta y$, and the volume between an assigned portion of the given surface and the plane of (x, y) will be found by taking the limit of the sum of a series of terms like $z \Delta x \Delta y$. Let V denote this volume, then

$$V = \iint z \, dx \, dy.$$

The equation to the surface gives z as a function of x and y; the limits of the integration must be taken so as to include all the elements of the proposed solid.

If we integrate first with respect to y, we sum up the columns which form a slice comprised between two planes perpendicular to the axis of x; thus the limits of integration with respect to y may be functions of x, and we shall obtain

$$\int z\,dy = f(x),$$

where $f(x)$ is in fact the area of the section of the solid considered made by a plane perpendicular to the axis of x at a distance x from the origin. Then finally

$$V = \int f(x)\,dx;$$

this coincides with the formula already given in Art. 188.

200. *Application to the Ellipsoid.*

Let it be required to find the volume of the eighth part of the ellipsoid determined by the equation

$$\frac{x^2}{a^2} + \frac{y^2}{b^2} + \frac{z^2}{c^2} = 1.$$

Here we have to find

$$\iint c\sqrt{\left(1 - \frac{x^2}{a^2} - \frac{y^2}{b^2}\right)}\,dx\,dy.$$

First integrate with respect to y, then the limits of y are 0 and Ll, that is, 0 and $b\sqrt{\left(1 - \frac{x^2}{a^2}\right)}$; we thus obtain the sum of all the columns which form the slice between the planes Lpl and Mqm. Now between the assigned limits

$$\int \sqrt{\left(1 - \frac{x^2}{a^2} - \frac{y^2}{b^2}\right)}\,dy = \frac{\pi b}{4}\left(1 - \frac{x^2}{a^2}\right);$$

thus
$$V = \int \frac{\pi}{4} bc\left(1 - \frac{x^2}{a^2}\right) dx.$$

The limits of x are 0 and a; we thus obtain the sum of

all the slices which are comprised in the solid *OABC*. Hence $V=\dfrac{\pi abc}{6}$.

201. Suppose the given surface to be determined by $xy=az$, and we require the volume bounded by the plane of (x, y), by the given surface, and by the four planes $x=x_1$, $x=x_2$, $y=y_1$, $y=y_2$. Here the volume is given by

$$V=\int_{x_1}^{x_2}\int_{y_1}^{y_2}\frac{xy}{a}\,dx\,dy$$

$$=\frac{1}{4a}(y_2^2-y_1^2)(x_2^2-x_1^2)$$

$$=\frac{1}{4a}(x_2-x_1)(y_2-y_1)\{x_1y_1+x_2y_2+x_1y_2+x_2y_1\}$$

$$=\frac{1}{4}(x_2-x_1)(y_2-y_1)(z_1+z_2+z_3+z_4),$$

where z_1, z_2, z_3, z_4 are the ordinates of the four corner points of the selected portion.

202. Find the volume comprised between the plane $z=0$ and the surfaces $xy=az$ and $(x-h)^2+(y-k)^2=c^2$.

Here we have to integrate $\iint\frac{xy}{a}\,dx\,dy$ between limits determined by $(x-h)^2+(y-k)^2=c^2$.

Now $\int y\,dy=\frac{y^2}{2}$, and the limits of y are

$$k-\sqrt{\{c^2-(x-h)^2\}} \text{ and } k+\sqrt{\{c^2-(x-h)^2\}}.$$

Thus we obtain

$$2k\sqrt{\{c^2-(x-h)^2\}}.$$

Thus finally the required volume

$$=\frac{2k}{a}\int x\sqrt{\{c^2-(x-h)^2\}}\,dx,$$

where the limits of x are $h-c$ and $h+c$.

And

$$\int x\sqrt{\{c^2-(x-h)^2\}}\,dx = \int (x-h)\sqrt{\{c^2-(x-h)^2\}}\,dx$$
$$+h\int\sqrt{\{c^2-(x-h)^2\}}\,dx.$$

Put $x-h=t$; thus we obtain

$$\int t\sqrt{(c^2-t^2)}\,dt + h\int\sqrt{(c^2-t^2)}\,dt.$$

The limits of t are $-c$ and $+c$; therefore the result is $\frac{hc^2\pi}{2}$; and the required volume is $\frac{hkc^2\pi}{a}$.

This result however assumes that xy is *positive* throughout the limits of the integration; that is, the circle determined by $(x-h)^2+(y-k)^2=c^2$ is supposed to lie entirely in the first quadrant or entirely in the third quadrant. If this condition be not fulfilled our result does not give the arithmetical value of the volume, but the balance arising from estimating some part of the volume as positive and some part as negative; for example, if h and k vanish our result vanishes.

Similarly in the result of the preceding article, it is assumed that xy is *positive* throughout the limits of the integration.

203. Instead of dividing a solid into columns standing on rectangular bases, so that $z\Delta x\Delta y$ is the volume of the column, we may divide it into columns standing upon the polar element of area; hence $zr\Delta\theta\Delta r$ is the volume of the column. Therefore for the volume V of a solid we have the formula

$$V=\iint z\,r\,d\theta\,dr.$$

From the equation to the surface z must be expressed as a function of r and θ.

For example, required the volume comprised between the plane $z=0$, and the surfaces $x^2+y^2=4az$ and $y^2=2cx-x^2$. Here $z=\frac{r^2}{4a}$; and the limits of r and θ must be such as to extend the integration over the whole area of the circle $y^2 = 2cx - x^2$. Let $r_1 = 2c\cos\theta$; then the required volume

$$= \int_{-\frac{1}{2}\pi}^{\frac{1}{2}\pi} \int_0^{r_1} \frac{r^3}{4a}\, d\theta\, dr$$

$$= \frac{c^4}{a} \int_{-\frac{1}{2}\pi}^{\frac{1}{2}\pi} \cos^4\theta\, d\theta$$

$$= \frac{2c^4}{a} \int_0^{\frac{1}{2}\pi} \cos^4\theta\, d\theta$$

$$= \frac{2c^4}{a} \cdot \frac{3}{4} \cdot \frac{1}{2} \cdot \frac{\pi}{2} \quad \text{(Art. 35)}$$

$$= \frac{3\pi c^4}{8a}.$$

204. Required the volume of the solid comprised between the plane of (x, y) and the surface whose equation is

$$z = ae^{-\frac{x^2+y^2}{c^2}}.$$

Here, since $x^2 + y^2 = r^2$,

$$V = a \iint e^{-\frac{r^2}{c^2}} r\, d\theta\, dr.$$

The surface extends to an infinite distance from the origin in every direction; thus the limits of θ are 0 and 2π, and those of r are 0 and ∞.

Now $$\int e^{-\frac{r^2}{c^2}} r dr = -\frac{e^{-\frac{r^2}{c^2}}}{2} c^2;$$

thus $$\int_0^\infty e^{-\frac{r^2}{c^2}} r dr = \frac{c^2}{2}.$$

And $$\int_0^{2\pi} d\theta = 2\pi.$$

Hence the required volume is $\pi a c^2$.

Formulæ involving Triple Integration.

205. In the figure to Art. 199, suppose we draw a series of planes perpendicular to the axis of z; let z be the distance of one plane from the origin and $z+\Delta z$ the distance of the next. These planes intercept from the column $pqPQ$ an elementary rectangular parallelopiped, the volume of which is $\Delta x \Delta y \Delta z$. The whole solid may be considered as the limit of the sum of such elements. Hence if V denote its volume,

$$V = \iiint dx\,dy\,dz.$$

206. Required the volume of a portion of the cylinder determined by the equation

$$x^2 + y^2 - 2ax = 0,$$

which is intercepted between the planes

$$z = x \tan \alpha \text{ and } z = x \tan \beta.$$

Here if y_1 stand for $\sqrt{(2ax - x^2)}$, we have

$$V = \int_0^{2a} \int_{-y_1}^{y_1} \int_{x \tan \alpha}^{x \tan \beta} dx\,dy\,dz$$

$$= \int_0^{2a} \int_{-y_1}^{y_1} (\tan \beta - \tan \alpha)\, x\,dx\,dy$$

$$= 2 (\tan \beta - \tan \alpha) \int_0^{2a} x \sqrt{(2ax - x^2)}\, dx$$

$$= 2 (\tan \beta - \tan \alpha) \frac{\pi a^3}{2}.$$

207. The polar element of plane area is, as we have seen in previous articles, $r\Delta\theta\,\Delta r$. Suppose this were to revolve round the initial line through an angle 2π, then a solid ring would be generated, of which the volume is $2\pi r \sin \theta\; r\Delta\theta\,\Delta r$,

since $2\pi r \sin\theta$ is the circumference of the circle described by the point whose polar co-ordinates are r and θ. Let ϕ denote the angle which the plane of the element in any position makes with the initial position of the plane, $\phi + \Delta\phi$ the angle which the plane in a consecutive position makes with the initial plane; then the part of the solid ring which is intercepted between the revolving plane in these two positions is to the whole ring in the same proportion as $\Delta\phi$ is to 2π. Hence the volume of this intercepted part is

$$r^2 \sin\theta\, \Delta\phi\, \Delta\theta\, \Delta r.$$

This is therefore an expression in polar co-ordinates for an element of any solid. Hence the volume of the whole solid may be found by taking the limit of the sum of such elements; that is, if V denote the required volume,

$$V = \iiint r^2 \sin\theta\, d\phi\, d\theta\, dr.$$

The limits of the integration must be so taken as to include in the integration all the elements of the proposed solid. The student will remember that r denotes the distance of any point from the origin, θ the angle which this distance makes with some fixed line through the origin, and ϕ the angle which the plane passing through this distance and the fixed line makes with some fixed plane passing through the fixed line.

208. Suppose, for example, that we apply the formula to find the volume of the eighth part of a sphere. Integrate with respect to r first; we have

$$\int r^2\, dr = \frac{r^3}{3}.$$

Suppose a the radius of the sphere, then the limits of r are 0 and a; thus

$$V = \iint \frac{a^3}{3} \sin\theta\, d\phi\, d\theta.$$

In thus integrating with respect to r, we collect all the elements like $r^2 \sin\theta\, \Delta\phi\, \Delta\theta\, \Delta r$ which compose a pyramidal solid, having its vertex at the centre of the sphere, and for its base the curvilinear element of spherical surface, which is denoted by $a^2 \sin\theta\, \Delta\phi\, \Delta\theta$.

Integrate next with respect to θ; we have

$$\int \sin\theta\, d\theta = -\cos\theta;$$

the limits of θ are 0 and $\frac{\pi}{2}$; thus

$$V = \int \frac{a^3}{3}\, d\phi.$$

In thus integrating with respect to θ, we collect all the pyramids similar to $\frac{a^3}{3} \sin\theta\, \Delta\phi\, \Delta\theta$ which form a wedge-shaped slice of the solid contained between the two planes through the fixed line corresponding to ϕ and $\phi + \Delta\phi$.

Lastly, integrate with respect to ϕ from 0 to $\frac{\pi}{2}$; thus

$$V = \frac{\pi a^3}{6}.$$

In this example the integrations may be performed in any order, and the student should examine and illustrate them.

209. A right cone has its vertex on the surface of a sphere, and its axis coincident with the diameter of the sphere passing through that point; find the volume common to the cone and the sphere.

Let a be the radius of the sphere; α the semi-vertical angle of the cone, V the required volume, then the polar equation to the sphere with the vertex of the cone as origin is $r = 2a \cos\theta$. Therefore

$$V = \int_0^{2\pi} \int_0^{\alpha} \int_0^{2a\cos\theta} r^2 \sin\theta\, d\phi\, d\theta\, dr.$$

210. The curve $r = a(1 + \cos\theta)$ revolves round the initial line, find the volume of the solid generated.

Here the required volume

$$=\int_0^{\pi}\int_0^{2\pi}\int_0^{a(1+\cos\theta)} r^2 \sin\theta \, d\theta \, d\phi \, dr$$

$$=\frac{2\pi a^3}{3}\int_0^{\pi}(1+\cos\theta)^3 \sin\theta \, d\theta.$$

It will be found that this $=\dfrac{8\pi a^3}{3}$.

EXAMPLES.

1. If the curve $y^2(x-4a)=ax(x-3a)$ revolve round the axis of x, the volume generated from $x=0$ to $x=3a$ is $\dfrac{\pi a^3}{2}(15-16\log 2)$.

2. A cycloid revolves round the tangent at the vertex; shew that the volume generated by the curve is $\pi^2 a^3$.

3. A cycloid revolves round its base; shew that the volume generated by the curve is $5\pi^2 a^3$.

4. The curve $y^2(2a-x)=x^3$ revolves round its asymptote; shew that the volume generated is $2\pi^2 a^3$.

5. The curve $xy^2=4a^2(2a-x)$ revolves round its asymptote; shew that the volume generated is $4\pi^2 a^3$.

6. Find the volume of the closed portion of the solid generated by the revolution of the curve $(y^2-b^2)^2=a^3x$ round the axis of y.

   Result. $\dfrac{256}{315}\dfrac{\pi b^9}{a^6}$.

7. Express the volume of a frustum of a sphere in terms of its height and the radii of its ends.

   Result. $\dfrac{\pi h}{6}\{h^2+3(r_1^2+r_2^2)\}$.

8. If the curve $y^2 = 2mx + nx^2$ revolve about the axis of x, find the volume of any frustum; and shew that it may be expressed either by

$$\frac{\pi h}{2}(b^2 + c^2 - \tfrac{1}{3}nh^2) \text{ or by } \pi h\left(r^2 + \frac{nh^2}{12}\right),$$

where h is the altitude of the frustum and b, c, r are the radii of its two ends and middle section. Deduce expressions for the volume of a cone and spheroid.

9. Find by integration the volume included between a right cone whose vertical angle is 60°, and a sphere of given radius touching it along a circle.

Result. $\dfrac{\pi r^3}{6}$.

10. If a paraboloid have its vertex in the base, and axis in the surface of a cylinder, the cylinder will be divided into parts which are as 3 : 5 by the surface of the paraboloid; the altitude and diameter of the base of the cylinder and the latus rectum of the paraboloid being all equal.

11. A paraboloid of revolution and a right cone have the same base, axis, and vertex, and a sphere is described upon this axis as diameter; shew that the volume intercepted between the paraboloid and cone bears the same ratio to the volume of the sphere that the latus rectum of the parabola bears to the diameter of the sphere.

12. Find the whole volume of the solid bounded by the surface of which the equation is

$$\frac{x^2}{a^2} + \frac{y^2}{b^2} + \frac{z^4}{c^4} = 1.$$

Result. $\dfrac{8\pi abc}{5}$.

13. Find the whole volume of the solid bounded by the surface of which the equation is

$$(x^2 + y^2 + z^2)^3 = 27a^3xyz.$$

Result. $\dfrac{9}{2}a^3$.

14. Find the volume of the solid formed by the revolution of the curve $(x^2+y^2)^2=a^2x^2+b^2y^2$ about the axis of x, supposing a greater than b. Shew what the result becomes when $a=b$.

Result. $\frac{\pi}{6}(2a^2+3b^2)\,a+\frac{\pi b^4}{2\sqrt{(a^2-b^2)}}\log\frac{a+\sqrt{(a^2-b^2)}}{b}$.

15. Determine the volume of the solid generated by the revolution of the curve $(x^2+y^2)^2=a^2x^2+b^2y^2$ about the axis of y, supposing a greater than b. Shew what the result becomes when $a=b$.

Result. $\frac{\pi}{6}(2b^2+3a^2)\,b+\frac{\pi a^4}{2\sqrt{(a^2-b^2)}}\sin^{-1}\frac{\sqrt{(a^2-b^2)}}{a}$.

16. Find the volume of the solid formed by the revolution of the curve $(y^2+x^2)^2=a^2(x^2-y^2)$ round the axis of x.

Result. $\frac{\pi a^3}{2}\left\{\frac{1}{\sqrt{2}}\log(1+\sqrt{2})-\frac{1}{3}\right\}$.

17. A paraboloid of revolution has its axis coincident with a diameter of a sphere, and its vertex outside the sphere; find the volume of the portion of the sphere outside the paraboloid.

Result. $\frac{\pi h^3}{6}$, where h is the distance of the two planes in which the curves of intersection of the surfaces are situated.

18. Find the volume cut off from the surface

$$\frac{z^2}{c}+\frac{y^2}{b}=2x$$

by a plane parallel to that of (y, z) at a distance a from it. *Result.* $\pi a^2\sqrt{(bc)}$.

19. A quadrant of an ellipse revolves round a tangent at the end of the minor axis of the ellipse; shew that

the volume included by the surface formed by the *curve* is

$$\frac{\pi ab^2}{6}(10 - 3\pi).$$

20. Find the volume enclosed by the surfaces defined by the equations

$$x^2 + y^2 = cz, \qquad x^2 + y^2 = ax, \qquad z = 0,$$

illustrating by figures the progress of the summation.

Result. $\frac{3\pi a^4}{32c}$.

21. If S be a closed surface, dS an element of S about a point P at a distance r from a fixed point O, and ϕ the angle which the normal at P drawn inwards makes with the radius vector OP, shew that the volume contained by the surface

$$= \tfrac{1}{3}\int r \cos\phi \, dS,$$

the summation being extended over the whole surface.

Taking the centre of an ellipsoid as the point O, apply this formula to find its volume, interpreting geometrically the steps of the integration.

22. Find the value of $\iiint x^2 \, dx \, dy \, dz$ over the volume of an ellipsoid. *Result.* $\frac{4\pi a^3 bc}{15}$.

23. Determine the limits of integration in order to obtain the volume contained between the plane of (x, y) and the surface whose equation is

$$Ax^2 + Bxy + Cy^2 - Dz - F = 0.$$

24. State the limits of the integration to be used in applying the formula $\iiint dx \, dy \, dz$ to find the volume of a closed surface of the second order whose equation is

$$ax^2 + by^2 + cz^2 + a'yz + b'xz + c'xy = 1.$$

25. State between what limits the integrations in

$$\iiint dx\,dy\,dz$$

must be performed, in order to obtain the volume contained between the conical surface whose equation is

$$z = a - \sqrt{(x^2 + y^2)},$$

and the planes whose equations are $x = z$ and $x = 0$; and find the volume by this or by any other method.

Result. $\dfrac{2a^3}{9}$.

26. State between what limits the integrations must be taken in order to find the volume of the solid contained between the two surfaces $cz = mx^2 + ny^2$ and $z = ax + by$; and shew that the volume is $\dfrac{\pi c^3}{8}$ when

$$m = n = a = b = 1.$$

27. A cavity is just large enough to allow of the complete revolution of a circular disc of radius c, whose centre describes a circle of the same radius c, while the plane of the disc is constantly parallel to a fixed plane, and perpendicular to that of the circle in which its centre moves. Shew that the volume of the cavity is

$$\frac{2c^3}{3}(3\pi + 8).$$

28. The axis of a right cone coincides with the generating line of a cylinder; the diameter of both cone and cylinder is equal to the common altitude; find the surface and volume of each part into which the cone is divided by the cylinder.

Results.

Surfaces, $\dfrac{4\pi\sqrt{5} - 3\sqrt{15}}{6}a^2$ and $\dfrac{2\pi\sqrt{5} + 3\sqrt{15}}{6}a^2$;

Volumes, $\frac{8\pi + 27\sqrt{3} - 64}{9} a^3$ and $\frac{64 - 27\sqrt{3} - 2\pi}{9} a^3$;

where a is the radius of the base of the cone or cylinder.

29. Find the volume of the cono-cuneus determined by

$$z^2 + \frac{a^2 y^2}{x^2} = c^2,$$

which is contained between the planes $x = 0$ and $x = a$.

Result. $\frac{\pi c^2 a}{2}$.

30. A conoid is generated by a straight line which passes through the axis of z and is perpendicular to it. Two sections are made by parallel planes, both planes being parallel to the axis of z. Shew that the volume of the conoid included between the planes is equal to the product of the distance of the planes into half the sum of the areas of the sections made by the planes.

CHAPTER IX.

DIFFERENTIATION OF AN INTEGRAL WITH RESPECT TO ANY QUANTITY WHICH IT MAY INVOLVE.

211. It is sometimes necessary to differentiate an integral with respect to some quantity which it involves; this question we shall now consider.

Required the differential coefficient of $\int_a^b \phi(x)\,dx$ with respect to b, supposing $\phi(x)$ not to contain b, and a to be independent of b.

Let
$$u = \int_a^b \phi(x)\,dx;$$

suppose b changed into $b + \Delta b$, in consequence of which u becomes $u + \Delta u$; thus

$$u + \Delta u = \int_a^{b+\Delta b} \phi(x)\,dx;$$

therefore
$$\Delta u = \int_a^{b+\Delta b} \phi(x)\,dx - \int_a^b \phi(x)\,dx$$
$$= \int_b^{b+\Delta b} \phi(x)\,dx.$$

Now, by Art. 40,

$$\int_b^{b+\Delta b} \phi(x)\,dx = \Delta b\,\phi(b + \theta\Delta b),$$

where θ is some proper fraction; thus

$$\frac{\Delta u}{\Delta b} = \phi(b + \theta\Delta b).$$

Let Δb and Δu diminish without limit; thus

$$\frac{du}{db} = \phi(b).$$

212. Similarly, if we differentiate u with respect to a, supposing $\phi(x)$ not to contain a, and b to be independent of a, we obtain

$$\frac{du}{da} = -\phi(a).$$

213. Suppose $\phi(x)$ to contain a quantity c, and let it be required to find the differential coefficient of $\int_a^b \phi(x)\,dx$ with respect to c, supposing a and b independent of c.

Instead of $\phi(x)$ it will be convenient to write $\phi(x, c)$, so that the presence of the quantity c may be more clearly indicated; denote the integral by u, thus

$$u = \int_a^b \phi(x, c)\,dx.$$

Suppose c changed into $c + \Delta c$, in consequence of which u becomes $u + \Delta u$; thus

$$u + \Delta u = \int_a^b \phi(x, c + \Delta c)\,dx;$$

therefore
$$\Delta u = \int_a^b \phi(x, c + \Delta c)\,dx - \int_a^b \phi(x, c)\,dx$$

$$= \int_a^b \{\phi(x, c + \Delta c) - \phi(x, c)\}\,dx;$$

thus
$$\frac{\Delta u}{\Delta c} = \int_a^b \frac{\phi(x, c + \Delta c) - \phi(x, c)}{\Delta c}\,dx.$$

Now by the nature of a differential coefficient we have

$$\frac{\phi(x, c + \Delta c) - \phi(x, c)}{\Delta c} = \frac{d\phi(x, c)}{dc} + \rho,$$

where ρ is a quantity which diminishes without limit when Δc does so. Thus we have

$$\frac{\Delta u}{\Delta c} = \int_a^b \frac{d\phi(x, c)}{dc} dx + \int_a^b \rho\, dx.$$

When Δc is diminished indefinitely, the second integral vanishes; for it is not greater than $(b - a)\rho'$, where ρ' is the greatest value ρ can have, and ρ' ultimately vanishes. Hence proceeding to the limit, we have

$$\frac{du}{dc} = \int_a^b \frac{d\phi(x, c)}{dc} dx.$$

214. It should be noticed that the preceding article supposes that neither a nor b is infinite; if, for example, b were infinite, we could not assert that $(b-a)\rho'$ would necessarily vanish in the limit.

215. We have shewn then in Art. 213 that

$$\frac{d}{dc}\int_a^b \phi(x, c)\, dx = \int_a^b \frac{d\phi(x, c)}{dc} dx \;\ldots\ldots\ldots\ldots (1).$$

We will point out a useful application of this equation. Suppose that $\psi(x, c)$ is the function of which $\phi(x, c)$ is the differential coefficient with respect to x, and that $\chi(x, c)$ is the function of which $\dfrac{d\phi(x, c)}{dc}$ is the differential coefficient with respect to x; thus (1) may be written

$$\frac{d\psi(b, c)}{dc} - \frac{d\psi(a, c)}{dc} = \chi(b, c) - \chi(a, c) \;\ldots\ldots\; (2),$$

let us suppose that b does not occur in $\phi(x, c)$, and that a is also independent of b; then (2) may be written

$$\frac{d\psi(b, c)}{dc} + C = \chi(b, c) \;\ldots\ldots\ldots\ldots\ldots\ldots\ldots (3),$$

where C denotes terms which are independent of b, that is, are constant with respect to b. Hence as b may have any value we please in (3), we may replace b by x, and write

$$\chi(x, c) = \frac{d\psi(x, c)}{dc} + C \text{..................} (4).$$

This equation may be applied to find $\chi(x, c)$; as the constant may be introduced if required, we may dispense with writing it, and put (4) in the form

$$\int \frac{d\phi(x, c)}{dc} dx = \frac{d}{dc} \int \phi(x, c)\, dx.$$

For example, let $\phi(x, c) = \frac{1}{1 + c^2x^2}$; then

$$\int \phi(x, c)\, dx = \int \frac{dx}{1 + c^2x^2} = \frac{1}{c} \tan^{-1} cx,$$

thus
$$\frac{d}{dc}\left(\frac{1}{c} \tan^{-1} cx\right) = \int \frac{d}{dc}\left(\frac{1}{1 + c^2x^2}\right) dx$$

$$= -\int \frac{2cx^2}{(1 + c^2x^2)^2} dx.$$

Thus from knowing the value of $\int \frac{dx}{1 + c^2x^2}$ we are able to deduce by differentiation the value of the more complex integral $\int \frac{2cx^2}{(1 + c^2x^2)^2} dx$.

216. Required the differential coefficient of $\int_a^b \phi(x, c)\, dx$ with respect to c when both b and a are functions of c. Denote the integral by u; then $\frac{du}{dc}$ consists of three terms, one arising from the fact that $\phi(x, c)$ contains c, one from the fact that b contains c, and one from the fact that a contains c.

Hence by the preceding articles,

$$\frac{du}{dc} = \int_a^b \frac{d\phi(x, c)}{dc} dx + \frac{du}{db}\frac{db}{dc} + \frac{du}{da}\frac{da}{dc}$$

$$= \int_a^b \frac{d\phi(x, c)}{dc} dx + \phi(b, c)\frac{db}{dc} - \phi(a, c)\frac{da}{dc}.$$

217. With the suppositions of the preceding article we may proceed to find $\frac{d^2u}{dc^2}$. By differentiating with respect to c the term $\int_a^b \frac{d\phi(x, c)}{dc} dx$ we obtain

$$\int_a^b \frac{d^2\phi(x, c)}{dc^2} dx + \frac{d\phi(b, c)}{dc}\frac{db}{dc} - \frac{d\phi(a, c)}{dc}\frac{da}{dc}.$$

From the other terms in $\frac{du}{dc}$ we obtain by differentiation

$$\phi(b, c)\frac{d^2b}{dc^2} + \frac{d\phi(b, c)}{db}\left(\frac{db}{dc}\right)^2 + \frac{d\phi(b, c)}{dc}\frac{db}{dc}$$

$$- \phi(a, c)\frac{d^2a}{dc^2} - \frac{d\phi(a, c)}{da}\left(\frac{da}{dc}\right)^2 - \frac{d\phi(a, c)}{dc}\frac{da}{dc}.$$

Thus $$\frac{d^2u}{dc^2} = \int_a^b \frac{d^2\phi(x, c)}{dc^2} dx$$

$$+ \phi(b, c)\frac{d^2b}{dc^2} + \frac{d\phi(b, c)}{db}\left(\frac{db}{dc}\right)^2 + 2\frac{d\phi(b, c)}{dc}\frac{db}{dc}$$

$$- \phi(a, c)\frac{d^2a}{dc^2} - \frac{d\phi(a, c)}{da}\left(\frac{da}{dc}\right)^2 - 2\frac{d\phi(a, c)}{dc}\frac{da}{dc}.$$

Similarly $\frac{d^3u}{dc^3}$ may be found and higher differential coefficients of u if required.

218. The following geometrical illustration may be given of Art. 216.

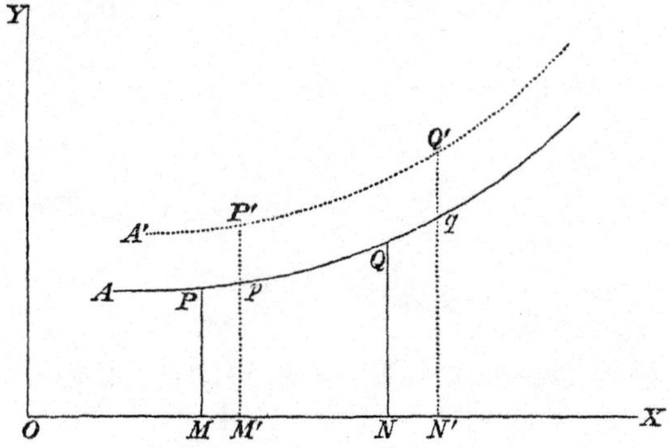

Let $y = \phi(x, c)$ be the equation to the curve APQ, and $y = \phi(x, c + \Delta c)$ the equation to the curve $A'P'Q'$.

Let $$OM = a, \qquad ON = b,$$
$$MM' = \Delta a, \qquad NN' = \Delta b.$$

Then u denotes the area $PMNQ$, and $u + \Delta u$ denotes the area $P'M'N'Q'$. Hence

$$\Delta u = P'pqQ' + QNN'q - PMM'p,$$

and $$\frac{\Delta u}{\Delta c} = \frac{P'pqQ'}{\Delta c} + \frac{QNN'q}{\Delta c} - \frac{PMM'p}{\Delta c}.$$

It may easily be seen that the limit of the first term is the limit of $\int_a^b \frac{\phi(x, c + \Delta c) - \phi(x, c)}{\Delta c}\, dx$, that the limit of the second term is the limit of $\phi(b, c)\frac{\Delta b}{\Delta c}$, and that the limit of the third term is the limit of $\phi(a, c)\frac{\Delta a}{\Delta c}$. This gives the result of Art. 216.

219. *Example.* Find a curve such that the area between the curve, the axis of x, and any ordinate, shall bear a constant ratio to the rectangle contained by that ordinate and the corresponding abscissa.

Suppose $\phi(x)$ the ordinate of the curve to the abscissa x; then $\int_0^c \phi(x)\,dx$ expresses the area between the curve, the axis of x, and the ordinate $\phi(c)$: hence by supposition we must have

$$\int_0^c \phi(x)\,dx = \frac{c\phi(c)}{n},$$

where n is some constant. This is to hold for all values of c; hence we may differentiate with respect to c; thus

$$\phi(c) = \frac{\phi(c)}{n} + \frac{c\phi'(c)}{n};$$

therefore $$c\phi'(c) = (n-1)\,\phi(c),$$

and $$\frac{\phi'(c)}{\phi(c)} = \frac{n-1}{c}.$$

By integration $\log\phi(c) = (n-1)\log c + \text{constant}$;

thus $$\phi(c) = Ac^{n-1},$$

and $$\phi(x) = Ax^{n-1},$$

which determines the required curve.

220. Find the form of $\phi(x)$, so that for all values of c

$$\frac{\int_0^c x\{\phi(x)\}^2 dx}{\int_0^c \{\phi(x)\}^2\,dx} = \frac{c}{n}.$$

By the supposition

$$\int_0^c x\{\phi(x)\}^2 dx = \frac{c}{n}\int_0^c \{\phi(x)\}^2\,dx.$$

Differentiate with respect to c; thus

$$c\{\phi(c)\}^2 = \frac{1}{n}\int_0^c \{\phi(x)\}^2\,dx + \frac{c}{n}\{\phi(c)\}^2;$$

thus $$c\left(1-\frac{1}{n}\right)\{\phi(c)\}^2 = \frac{1}{n}\int_0^c \{\phi(x)\}^2\,dx.$$

Differentiate again with respect to c;

thus $$\left(1-\frac{1}{n}\right)\{\phi(c)\}^2 + 2c\left(1-\frac{1}{n}\right)\phi(c)\,\phi'(c) = \frac{\{\phi(c)\}^2}{n};$$

hence $$\left(1-\frac{2}{n}\right)\phi(c) + 2c\left(1-\frac{1}{n}\right)\phi'(c) = 0;$$

therefore $$\frac{\phi'(c)}{\phi(c)} = \frac{2-n}{2(n-1)}\,\frac{1}{c}.$$

Integrate; thus

$$\log\phi(c) = \frac{2-n}{2(n-1)}\log c + \text{constant};$$

therefore $$\phi(c) = Ac^{\frac{2-n}{2(n-1)}},$$

where A is some constant; thus we have finally

$$\phi(x) = Ax^{\frac{2-n}{2(n-1)}}.$$

This is the solution of a problem in Analytical Statics, which may be enunciated thus. The distance of the centre of gravity of a segment of a solid of revolution from the vertex is always $\frac{1}{n}$th part of the height of the segment; find the generating curve. The required equation is $y = \phi(x)$.

221. Find the form of $\phi(x)$ so that the integral $\int_0^c \frac{\phi(x)\,dx}{\sqrt{(c-x)}}$ may be independent of c.

Denote the integral by u, and suppose $x = cz$; thus

$$u = \int_0^c \frac{\phi(x)\,dx}{\sqrt{(c-x)}} = \int_0^1 \frac{\sqrt{c}\,\phi(cz)\,dz}{\sqrt{(1-z)}}.$$

Since u is to be independent of c, the differential coefficient of u with respect to c must vanish. Now

$$\frac{du}{dc}=\int_0^1 \frac{\dfrac{\phi(cz)}{2\sqrt{c}}+z\sqrt{c}\,\phi'(cz)}{\sqrt{(1-z)}}\,dz$$

$$=\int_0^c \frac{\phi(x)+2x\phi'(x)}{2c\sqrt{(c-x)}}\,dx.$$

This last integral then must vanish whatever c may be; hence we must have

$$\phi(x)+2x\phi'(x)=0;$$

therefore $$\frac{\phi'(x)}{\phi(x)}=-\frac{1}{2x};$$

therefore $$\log\phi(x)=-\tfrac{1}{2}\log x+\text{constant},$$

therefore $$\phi(x)=\frac{A}{\sqrt{x}}.$$

This is the solution of a problem in Dynamics, which may be thus enunciated. Find a curve, such that the time of falling down an arc of the curve from *any* point to the lowest point may be the same. If s denote the arc of the curve measured from the lowest point, x the horizontal abscissa of the extremity of s, then we have

$$\frac{ds}{dx}=\phi(x)\quad\text{and}\quad s=2A\sqrt{x};$$

so that the curve is a cycloid (Art. 72).

MISCELLANEOUS EXAMPLES.

1. If the straight line $SP_1P_2P_3$ meet three successive revolutions of an equiangular spiral, whose equation is $r=a^\theta$, in the points P_1, P_2, P_3, find the area included between P_1P_2, P_2P_3, and the two curve lines P_1P_2, P_2P_3.

Result. $\dfrac{1}{4\log_e a}(P_3P_1)^2$.

2. Find the area of the curve $y^4 - axy^2 + x^4 = 0$.

Result. $\dfrac{\pi a^2 \sqrt{2}}{16}$.

3. Find the area of the curve $x^{2n} + y^{2n} = a^2 (xy)^{n-1}$, where n is a positive integer.

Result. If n is an *even* integer $\dfrac{a^2\pi}{2n}$; if n is an *odd* integer $\dfrac{a^2\pi}{n}$.

4. A string the length of which is equal to the perimeter of an oval is wound completely round the oval, and an involute is formed by unwinding the string, beginning at any point; shew that when the length of the involute is a maximum or a minimum the length of the string is equal to the perimeter of the circle of curvature at the point from which the unwinding begins.

5. Find the portion of the cylinder $x^2 + y^2 - rx = 0$ intercepted between the planes

$$ax + by + cz = 0 \text{ and } a'x + by + cz = 0.$$

Result. $\dfrac{\pi (a' - a) r^3}{8c}$.

6. Find the volume of the solid bounded by the paraboloid $y^2 + z^2 = 4a(x + a)$ and the sphere $x^2 + y^2 + z^2 = c^2$, supposing c greater than a.

Result. $2\pi a \left(c^2 - \dfrac{a^2}{3}\right)$.

CHAPTER X.

ELLIPTIC INTEGRALS.

222. THE integrals $\int \frac{d\theta}{\sqrt{(1-c^2\sin^2\theta)}}$, $\int \sqrt{(1-c^2\sin^2\theta)}\,d\theta$, and $\int \frac{d\theta}{(1+a\sin^2\theta)\sqrt{(1-c^2\sin^2\theta)}}$, are called *elliptic functions* or *elliptic integrals* of the first, second, and third order respectively; the first is denoted by $F(c, \theta)$, the second by $E(c, \theta)$, and the third by $\Pi(c, a, \theta)$. The integrals are all supposed to be taken between the limits 0 and θ, so that they vanish with θ; θ is called the *amplitude* of the function. The constant c is supposed less than unity; it is called the *modulus* of the function. The constant a, which occurs in the function of the third order, is called the *parameter*. When the integrals are taken between the limits 0 and $\frac{\pi}{2}$, they are called *complete* functions; that is, the *amplitude* of a complete function is $\frac{\pi}{2}$.

223. The second elliptic integral expresses the length of a portion of the arc of an ellipse measured from the end of the minor axis, the excentricity of the ellipse being the *modulus* of the function. From this circumstance, and from the fact that the three integrals are connected by remarkable properties, the name *elliptic integrals* has been derived.

224. The subject of *elliptic integrals* is very extensive; we shall merely give a few of the simpler results, and refer the student for fuller investigations to Hymers's *Integral Calculus*, or to the writings of Legendre, Jacobi and Abel.

225. If θ and ϕ are connected by the equation

$$F(c, \theta) + F(c, \phi) = F(c, \mu),$$

where μ is a constant; then will

$$\cos\theta\cos\phi - \sin\theta\sin\phi\sqrt{(1 - c^2\sin^2\mu)} = \cos\mu.$$

Consider θ and ϕ as functions of a new variable t, and differentiate the given equation; thus

$$\frac{1}{\sqrt{(1 - c^2\sin^2\theta)}}\frac{d\theta}{dt} + \frac{1}{\sqrt{(1 - c^2\sin^2\phi)}}\frac{d\phi}{dt} = 0 \ldots\ldots (1).$$

Now as t is a new arbitrary variable, we are at liberty to assume

$$\frac{d\theta}{dt} = \sqrt{(1 - c^2\sin^2\theta)},$$

thus from the equation (1)

$$\frac{d\phi}{dt} = -\sqrt{(1 - c^2\sin^2\phi)}.$$

Square these two equations and differentiate; thus

$$\frac{d^2\theta}{dt^2} = -c^2\sin\theta\cos\theta, \qquad \frac{d^2\phi}{dt^2} = -c^2\sin\phi\cos\phi;$$

therefore $$\frac{d^2(\theta \pm \phi)}{dt^2} = -\frac{c^2}{2}(\sin 2\theta \pm \sin 2\phi).$$

Let $\theta + \phi = \psi$ and $\theta - \phi = \chi$; thus

$$\frac{d^2\psi}{dt^2} = -c^2\sin\psi\cos\chi, \qquad \frac{d^2\chi}{dt^2} = -c^2\sin\chi\cos\psi.$$

Also $$\frac{d\psi}{dt}\frac{d\chi}{dt} = \left(\frac{d\theta}{dt}\right)^2 - \left(\frac{d\phi}{dt}\right)^2 = -c^2\sin\psi\sin\chi;$$

therefore $$\frac{\frac{d^2\psi}{dt^2}}{\frac{d\psi}{dt}\frac{d\chi}{dt}} = \cot\chi, \quad \frac{\frac{d^2\chi}{dt^2}}{\frac{d\psi}{dt}\frac{d\chi}{dt}} = \cot\psi;$$

therefore

$$\frac{d}{dt}\left(\log\frac{d\psi}{dt}\right) = \frac{d}{dt}\log\sin\chi, \quad \frac{d}{dt}\left(\log\frac{d\chi}{dt}\right) = \frac{d}{dt}\log\sin\psi;$$

therefore $$\log\frac{d\psi}{dt} = \log\sin\chi + \text{constant},$$

therefore $$\frac{d\psi}{dt} = A\sin\chi$$

and similarly, $$\frac{d\chi}{dt} = B\sin\psi \quad \text{.................. (2)},$$

where A and B are constants.

Hence $$A\sin\chi\frac{d\chi}{dt} = B\sin\psi\frac{d\psi}{dt},$$

therefore $$A\cos\chi = B\cos\psi + C \text{............... (3)}.$$

Now from the original given equation we see that if $\phi = 0$

$$F(c, \theta) = F(c, \mu);$$

therefore then $\theta = \mu$ and $\chi = \psi = \mu$;

thus from (3) $(A - B)\cos\mu = C$;

thus $A\cos(\theta - \phi) = B\cos(\theta + \phi) + (A - B)\cos\mu$;

therefore

$$(A - B)\cos\theta\cos\phi + (A + B)\sin\theta\sin\phi = (A - B)\cos\mu \text{...(4)}.$$

In (2) put for $\frac{d\psi}{dt}$ its value

$$\sqrt{(1 - c^2\sin^2\theta)} - \sqrt{(1 - c^2\sin^2\phi)},$$

and for $\frac{d\chi}{dt}$ its value

$$\sqrt{(1-c^2\sin^2\theta)}+\sqrt{(1-c^2\sin^2\phi)},$$

and then suppose $\phi=0$; thus

$$\sqrt{(1-c^2\sin^2\mu)}-1=A\sin\mu,$$

and
$$\sqrt{(1-c^2\sin^2\mu)}+1=B\sin\mu.$$

Substitute for $A-B$ and $A+B$ in (4);

thus
$$\cos\theta\cos\phi-\sin\theta\sin\phi\sqrt{(1-c^2\sin^2\mu)}=\cos\mu.$$

226. The relation just found may be put in a different form. Clear the equation of radicals; thus

$$(\cos\theta\cos\phi-\cos\mu)^2=(1-c^2\sin^2\mu)\sin^2\theta\sin^2\phi;$$

therefore

$$\cos^2\theta+\cos^2\phi+\cos^2\mu-2\cos\theta\cos\phi\cos\mu$$
$$=1-c^2\sin^2\mu\sin^2\theta\sin^2\phi.$$

Add $\cos^2\phi\cos^2\mu$ to both sides and transpose; thus

$$(\cos\theta-\cos\phi\cos\mu)^2$$
$$=1-\cos^2\phi-\cos^2\mu+\cos^2\phi\cos^2\mu-c^2\sin^2\mu\sin^2\theta\sin^2\phi$$
$$=\sin^2\phi\sin^2\mu(1-c^2\sin^2\theta);$$

therefore
$$\cos\theta=\cos\phi\cos\mu+\sin\phi\sin\mu\sqrt{(1-c^2\sin^2\theta)}.$$

The positive sign of the radical is taken, because when $\theta=0$, we must have $\phi=\mu$.

227. We shall now shew how an elliptic function of the first order may be connected with another having a different modulus.

Let $F(c,\theta)$ denote the function; assume

$$\tan\theta=\frac{\sin 2\phi}{c+\cos 2\phi};$$

therefore $$\frac{1}{\cos^2\theta}\frac{d\theta}{d\phi} = \frac{2(1+c\cos 2\phi)}{(c+\cos 2\phi)^2},$$

therefore $$\frac{d\theta}{d\phi} = \frac{2(1+c\cos 2\phi)}{1+2c\cos 2\phi + c^2}.$$

And $$1 - c^2\sin^2\theta = 1 - \frac{c^2\sin^2 2\phi}{1+2c\cos 2\phi + c^2}$$

$$= \frac{1+2c\cos 2\phi + c^2\cos^2 2\phi}{1+2c\cos 2\phi + c^2};$$

therefore

$$\int\frac{d\theta}{\sqrt{(1-c^2\sin^2\theta)}} = \int\frac{2(1+c\cos 2\phi)}{1+2c\cos 2\phi + c^2}\cdot\frac{\sqrt{(1+2c\cos 2\phi + c^2)}}{1+c\cos 2\phi}\,d\phi$$

$$= 2\int\frac{d\phi}{\sqrt{(1+2c\cos 2\phi + c^2)}} = \frac{2}{1+c}\int\frac{d\phi}{\sqrt{\left\{1-\frac{4c}{(1+c)^2}\sin^2\phi\right\}}}.$$

No constant is added, because ϕ vanishes with θ. Thus $F(c, \theta) = \frac{2}{1+c}F(c_1, \phi)$, where

$$c_1^2 = \frac{4c}{(1+c)^2} \text{ and } \tan\theta = \frac{\sin 2\phi}{c+\cos 2\phi}.$$

The last relation may be written thus,

$$c\sin\theta = \sin(2\phi - \theta).$$

We may notice that c_1 is greater than c, for

$$\frac{c_1^2}{c^2} = \frac{4}{c(1+c)^2},$$

and since c is less than unity, 4 is greater than $c(1+c)^2$.

If $\phi = \frac{\pi}{2}$, then $\theta = \pi$; thus

$$\frac{2}{1+c}F\left(c_1, \frac{\pi}{2}\right) = F(c, \pi) = 2F\left(c, \frac{\pi}{2}\right).$$

228. We will give one more proposition in this subject, by establishing a relation among Elliptic Functions of the second order, analogous to that proved in Art. 225 for functions of the first order.

If $\cos\theta\cos\phi - \sin\theta\sin\phi\sqrt{(1-c^2\sin^2\mu)} = \cos\mu$,

then will

$$E(c,\theta) + E(c,\phi) - E(c,\mu) = c^2\sin\theta\sin\phi\sin\mu.$$

By virtue of the given equation connecting the amplitudes, ϕ is a function of θ; thus we may assume

$$E(c,\theta) + E(c,\phi) - E(c,\mu) = f(\theta).$$

Differentiate; thus

$$f'(\theta) = \sqrt{(1-c^2\sin^2\theta)} + \sqrt{(1-c^2\sin^2\phi)}\,\frac{d\phi}{d\theta}$$

$$= \frac{\cos\theta - \cos\phi\cos\mu}{\sin\phi\sin\mu} + \frac{\cos\phi - \cos\theta\cos\mu}{\sin\theta\sin\mu}\,\frac{d\phi}{d\theta}$$

(by Art. 226),

$$= \frac{d\{\sin^2\theta + \sin^2\phi + 2\cos\theta\cos\phi\cos\mu\}}{d\theta} \times \frac{1}{2\sin\theta\sin\phi\sin\mu}.$$

But $\sin^2\theta + \sin^2\phi + 2\cos\theta\cos\phi\cos\mu$

$$= 1 + \cos^2\mu + c^2\sin^2\theta\sin^2\phi\sin^2\mu;$$

thus $$f'(\theta) = c^2\sin\mu\,\frac{d(\sin\theta\sin\phi)}{d\theta}.$$

Therefore, by integration

$$f(\theta) = c^2\sin\theta\sin\phi\sin\mu.$$

No constant is added, because $f(\theta)$ obviously vanishes with θ.

MISCELLANEOUS EXAMPLES.

1. Find the whole volume of the solid bounded by the surface of which the equation is

$$z^2 = \frac{2axy}{\sqrt{(x^2+y^2)}} - (x^2+y^2).$$

Result. $\frac{\pi a^3}{6}$; supposing the radical restricted to the positive sign.

2. Find the whole volume of the solid bounded by the surface of which the equation is

$$\left(\frac{x}{a}\right)^{\frac{2}{3}} + \left(\frac{y}{b}\right)^{\frac{2}{3}} + \left(\frac{z}{c}\right)^{\frac{2}{3}} = 1.$$

Result. $\frac{4\pi abc}{35}$.

3. Prove that the volume of that portion of the solid bounded by the surface whose equation is

$$x^2z + ay^2 = z(a^2 - z^2),$$

which lies on the positive side of the plane of xy is $\frac{8\pi a^3}{21}$.

4. Find the value of $\int \frac{dS}{r^n}$, where dS denotes the element of the surface of a sphere, and r the distance of this element from a fixed point without the sphere; the integration being extended over the whole surface of the sphere.

Result. $\frac{2\pi a}{c(n-2)}\left\{\frac{1}{(c-a)^{n-2}} - \frac{1}{(c+a)^{n-2}}\right\}$; where a is the radius of the sphere, and c the distance of the fixed point from the centre of the sphere.

5. A cylinder is constructed on a single loop of the curve $r = a\cos n\theta$ having its generating lines perpendicular to the plane of this curve; determine the area of the portion of the surface of the sphere $x^2+y^2+z^2=a^2$

which the cylinder intercepts; determine also the volume of the cylinder which the sphere intercepts.

Results. The area $= \frac{4a^2}{n}\left(\frac{\pi}{2} - 1\right);$

the volume $= \frac{4a^3}{3n}\left(\frac{\pi}{2} - \frac{2}{3}\right).$

6. Find the volume of the solid generated by the revolution of the closed part of the curve

$$x^3 - 3axy + y^3 = 0$$

round the line $x + y = 0$. *Result.* $\frac{8\pi^2 a^3}{3\sqrt{6}}$.

7. If the axes of two equal circular cylinders of radius a intersect at an angle β, the volume common to both is $\frac{16}{3}\frac{a^3}{\sin\beta}$; and the surface of each intercepted by the other is $\frac{8a^2}{\sin\beta}$.

8. The centre of a variable circle moves along the arc of a fixed circle; its plane is normal to the fixed circle, and its radius equal to the distance of its centre from a fixed diameter; find the volume generated; and if the solid so formed revolve round the fixed diameter, shew that the volume swept through is to the volume of the solid as 5 to 2.

9. The centre of a regular hexagon moves along a diameter of a given circle (radius $= a$), the plane of the hexagon being perpendicular to this diameter and its magnitude varying in such a manner that one of its diagonals always coincides with a chord of the circle; shew that the volume of the solid generated is $2\sqrt{3}\,a^3$. Shew also that the surface of the solid is

$$a^2 (2\pi + 3\sqrt{3}).$$

10. Prove that

$$\int_0^{\frac{a}{2}} \frac{dx}{\sqrt{(2ax - x^2)}\sqrt{(a^2 - x^2)}} = \frac{2}{3a} F\left(c, \frac{\pi}{2}\right), \text{ where } c = \frac{1}{3}.$$

CHAPTER XI.

CHANGE OF THE VARIABLES IN A MULTIPLE INTEGRAL.

229. WE have seen in Art. 62 that the double integral $\int_a^b \int_\alpha^\beta \phi(x, y)\, dx\, dy$ is equal to $\int_\alpha^\beta \int_a^b \phi(x, y)\, dy\, dx$ when the *limits are constant*, that is, a change in the *order* of integration produces no change in the limits for the two integrations. But when the limits of the first integration are functions of the other variable, this statement no longer holds, as we have seen in several examples in the seventh and eighth chapters. We give here a few additional examples.

230. Change the order of integration in

$$\int_0^a \int_0^{\sqrt{(a^2-x^2)}} \phi(x, y)\, dx\, dy.$$

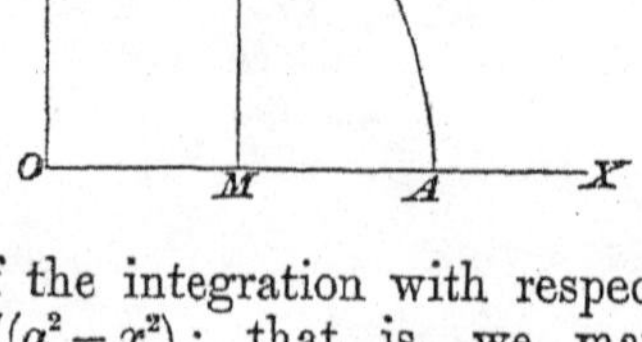

The limits of the integration with respect to y here are $y=0$ and $y=\sqrt{(a^2-x^2)}$; that is, we may consider the integral extending from the axis of x to the boundary of a circle, having its centre at the origin, and radius equal to a. Then the integration with respect to x extends from the axis of y to the extreme point A of the quadrant. Thus if we consider $z = \phi(x, y)$ as the equation to a surface, the above double integral represents the volume of that solid which is contained between the surface, the plane of (x, y), and a line moving perpendicularly to this plane round the boundary $OAPBO$.

It is then obvious from the figure that if the integration with respect to x is performed first, the limits will be $x = 0$ and $x = \sqrt{(a^2 - y^2)}$, and then the limits for y will be $y = 0$ and $y = a$. Thus the transformed integral is

$$\int_0^a \int_0^{\sqrt{(a^2-y^2)}} \phi(x, y)\, dy\, dx.$$

231. Change the order of integration in

$$\int_0^{\frac{\pi}{2}} \int_0^{2a\cos\theta} \phi(r, \theta)\, r\, d\theta\, dr.$$

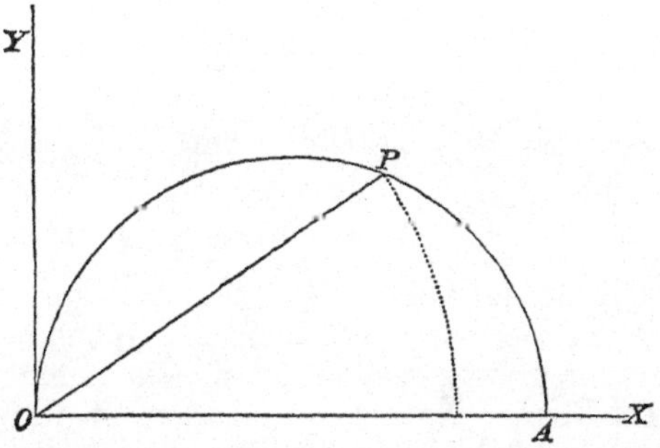

Let $OA = 2a$, and describe a semicircle on OA as diameter. Let $POX = \theta$, then $OP = 2a\cos\theta$. Thus the double integral may be considered as the limit of a summation of values of $\phi(r, \theta)\, r\, \Delta\theta\, \Delta r$ over all the area of the semicircle. Hence when the order is changed we must integrate for θ from 0 to $\cos^{-1}\frac{r}{2a}$, and for r from 0 to $2a$.

Thus the transformed integral is

$$\int_0^{2a}\int_0^{\cos^{-1}\frac{r}{2a}} \phi(r,\theta)\, r\, dr\, d\theta.$$

232. Change the order of integration in

$$\int_0^{2a}\int_{\frac{x^2}{4a}}^{3a-x} \phi(x,y)\, dx\, dy.$$

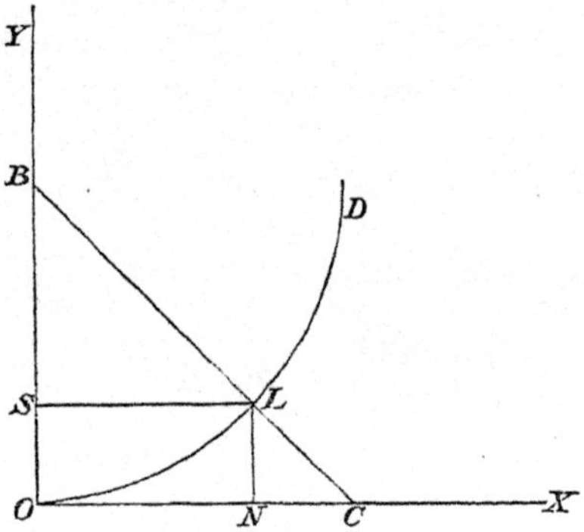

The integration for y is taken from $y=\frac{x^2}{4a}$ to $y=3a-x$. The equation $y=\frac{x^2}{4a}$ belongs to a parabola OLD, and $y=3a-x$ to a straight line BLC, which passes through L, the extremity of the latus rectum of the parabola.

Thus the integration may be considered as extending over the area $OLBSO$. Now let the order of integration be changed; we shall have to consider separately the spaces OLS and BLS. For the space OLS we must integrate from $x=0$ to $x=2\sqrt{(ay)}$, and then from $y=0$ to $y=a$; and for the space BLS we must integrate from $x=0$ to $x=3a-y$, and then from $y=a$ to $y=3a$. Thus the transformed integral is

$$\int_0^a\int_0^{2\sqrt{(ay)}} \phi(x,y)\, dy\, dx + \int_a^{3a}\int_0^{3a-y} \phi(x,y)\, dy\, dx.$$

233. Change the order of integration in

$$\int_0^1 \int_x^{x(2-x)} \phi(x, y)\, dx\, dy.$$

Here the integration with respect to y is taken from $y=x$ to $y=x(2-x)$. The equation $y=x$ represents a straight line, and $y=x(2-x)$ represents a parabola. The reader will find on examining a figure, that the transformed integral is

$$\int_0^1 \int_{1-\sqrt{(1-y)}}^{y} \phi(x, y)\, dy\, dx.$$

234. Change the order of integration in

$$\int_0^a \int_{\sqrt{(a^2-x^2)}}^{x+2a} \phi(x, y)\, dx\, dy.$$

Here the integration with respect to y is taken from $y=\sqrt{(a^2-x^2)}$ to $y=x+2a$. The equation $y=\sqrt{(a^2-x^2)}$ represents a circle, and $y=x+2a$ represents a straight line. The reader will find on examining a figure, that when the integration with respect to x is performed first, the integral must be separated into three portions; the transformed integral is

$$\int_0^a \int_{\sqrt{(a^2-y^2)}}^{a} \phi(x, y)\, dy\, dx + \int_a^{2a} \int_0^a \phi(x, y)\, dy\, dx$$
$$+ \int_{2a}^{3a} \int_{y-2a}^{a} \phi(x, y)\, dy\, dx.$$

235. Change the order of integration in

$$\int_0^a \int_0^{\frac{b}{b+x}} \phi(x, y)\, dx\, dy.$$

Here the integration with respect to y is taken from $y=0$ to $y=\frac{b}{b+x}$. The equation $y=\frac{b}{b+x}$ represents an hyperbola; let BDE be this hyperbola, and let $OA=a$. Then the integration may be considered as extending over the

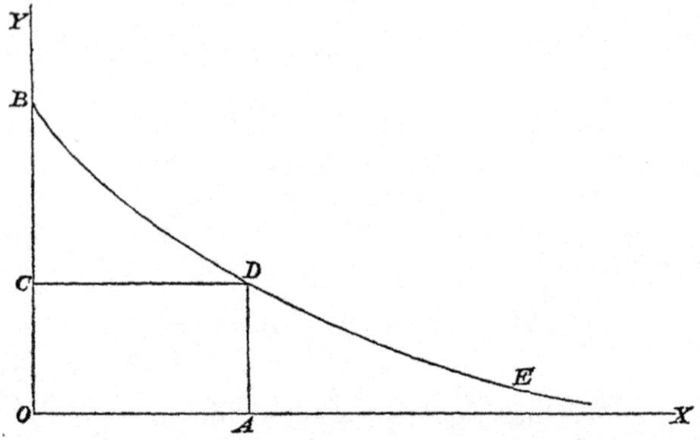

space $OBDA$. Let the order of the integration be changed; we shall then have to consider separately the spaces $OADC$ and CDB. For the space $OADC$ we must integrate from $x=0$ to $x=a$, and then from $y=0$ to $y=\frac{b}{b+a}$. For the space CDB we must integrate from $x=0$ to $x=\frac{b(1-y)}{y}$, and then from $y=\frac{b}{b+a}$ to $y=1$. Thus the transformed integral is

$$\int_0^{\frac{b}{b+a}}\int_0^a \phi(x, y)\, dy\, dx + \int_{\frac{b}{b+a}}^1 \int_0^{\frac{b(1-y)}{y}} \phi(x, y)\, dy\, dx.$$

236. Change the order of integration in

$$\int_0^h \int_{\lambda x}^{c-\mu x} \phi(x, y)\, dx\, dy,$$

where $h=\frac{c}{\lambda+\mu}$. The transformed integral is

$$\int_0^{\lambda h}\int_0^{\frac{y}{\lambda}} \phi(x, y)\, dy\, dx + \int_{\lambda h}^c \int_0^{\frac{c-y}{\mu}} \phi(x, y)\, dy\, dx.$$

237. Change the order of integration in

$$\int_0^a \int_0^x \int_0^y \phi(x, y, z)\, dx\, dy\, dz.$$

The integration here may be considered to be extended throughout a pyramid, the bounding planes of which are given by the equations

$$z = 0, \quad z = y, \quad y = x, \quad x = a.$$

The integral may be transformed in different ways, and thus we obtain

$$\int_0^a \int_y^a \int_0^y \phi(x, y, z)\, dy\, dx\, dz,$$

or

$$\int_0^a \int_0^y \int_y^a \phi(x, y, z)\, dy\, dz\, dx,$$

or

$$\int_0^a \int_z^a \int_y^a \phi(x, y, z)\, dz\, dy\, dx,$$

or

$$\int_0^a \int_0^x \int_z^x \phi(x, y, z)\, dx\, dz\, dy,$$

or

$$\int_0^a \int_z^a \int_z^x \phi(x, y, z)\, dz\, dx\, dy.$$

These transformations may be verified by putting for $\phi(x, y, z)$ some simple function, so that the integrals can be actually obtained; for example, if we replace $\phi(x, y, z)$ by unity, we find $\frac{a^3}{6}$ as the value of any one of the six forms.

238. These examples will sufficiently illustrate the subject; it is impossible to lay down any simple rules for the discovery of the limits of the transformed integral. It is not absolutely necessary to draw figures as we have done, for the figures convey no information which could not be obtained by reflection on the different values which the variables must have, in order to make the integration extend over the range indicated by the given limits. But the figures materially assist in arriving speedily and correctly at the result.

We now proceed to the problem which is the main object of the present chapter, namely, the change of the variables in a *multiple* integral. We begin with the case of a *double* integral.

239. The problem to be solved is the following. Required to transform the double integral $\iint V\,dx\,dy$, where V is a function of x and y, into another double integral in which the variables are u and v, the old and new variables being connected by the equations

$$\phi_1(x, y, u, v) = 0, \qquad \phi_2(x, y, u, v) = 0 \;\ldots\ldots(1).$$

We suppose that the original integral is to be taken between known limits of y and x; as we integrate with respect to y first, the limits of y may be functions of x. Of course while integrating with respect to y we regard x as constant.

We first transform the integral with respect to y into an integral with respect to v. This is theoretically very simple; from equations (1) eliminate u and obtain y as a function of x and v, say

$$y = \psi(x, v) \ldots\ldots\ldots\ldots\ldots\ldots\ldots\ldots(2),$$

from which we get

$$dy = \psi'(x, v)\,dv,$$

where $\psi'(x, v)$ means the differential coefficient of $\psi(x, v)$ with respect to v.

Substitute then for y and dy in $\int V\,dy$, and we obtain $\int V_1\psi'(x, v)\,dv$, where V_1 is what V becomes when we put for y its value in V. Hence the original double integral becomes

$$\iint V_1\psi'(x, v)\,dx\,dv.$$

Thus we have removed y and taken v instead. As the limiting values of y between which we had originally to integrate are known, we shall from (2) know the limiting values of v, between which we ought to integrate. It will be observed, that in finding $\frac{dy}{dv}$ from (2), we supposed x constant; this we do because, as already remarked, when we integrate the proposed expression with respect to y we must consider x constant.

The next step is to change the *order* of the above integrations with respect to x and v, that is, to perform the integration with respect to x *first*. This is a subject which we have already examined; all we have to do is to determine the *new limits* properly. Thus supposing this point settled, we have changed the original expression into

$$\iint V_1 \psi'(x, v)\, dv\, dx.$$

It remains to remove x from this expression and replace it by u. We proceed precisely as before. From equations (1) eliminate y, and obtain x as a function of v and u, say

$$x = \chi(v, u) \dots\dots\dots\dots\dots\dots (3),$$

from which we get

$$dx = \chi'(v, u)\, du,$$

where $\chi'(v, u)$ means the differential coefficient of $\chi(v, u)$ with respect to u.

Substitute then for x and dx, and the double integral becomes

$$\iint V' \psi'(x, v)\, \chi'(v, u)\, dv\, du,$$

where V' is what V_1 becomes when we put for x its value in V_1. Thus the double integral now contains only u and v, since for the x which occurs in $\psi'(x, v)$ we suppose its value substituted, namely, $\chi(v, u)$. Moreover since the limits between which the integration with respect to x was to be taken have been already settled, we know the limits between which the integration with respect to u must be taken.

We have thus given the complete *theoretical* solution of the problem; it only remains to add a *practical* method for determining $\psi'(x, v)$ and $\chi'(v, u)$; to this we proceed.

We observe that $\psi'(x, v)$ or $\frac{dy}{dv}$ is to be found from equations (1) by eliminating u, considering x constant; the following is exactly equivalent; from (1)

$$\frac{d\phi_1}{dy}\frac{dy}{dv}+\frac{d\phi_1}{du}\frac{du}{dv}+\frac{d\phi_1}{dv}=0,$$

$$\frac{d\phi_2}{dy}\frac{dy}{dv}+\frac{d\phi_2}{du}\frac{du}{dv}+\frac{d\phi_2}{dv}=0.$$

Eliminate $\frac{du}{dv}$; thus

$$\frac{\frac{d\phi_1}{dy}\frac{dy}{dv}+\frac{d\phi_1}{dv}}{\frac{d\phi_1}{du}}=\frac{\frac{d\phi_2}{dy}\frac{dy}{dv}+\frac{d\phi_2}{dv}}{\frac{d\phi_2}{du}},$$

therefore
$$\frac{dy}{dv}=\frac{\frac{d\phi_1}{dv}\frac{d\phi_2}{du}-\frac{d\phi_1}{du}\frac{d\phi_2}{dv}}{\frac{d\phi_1}{du}\frac{d\phi_2}{dy}-\frac{d\phi_1}{dy}\frac{d\phi_2}{du}}.$$

This then is an equivalent for $\psi'(x, v)$, supposing that after the differentiations are performed we put for y and u their values in terms of x and v from (1).

Again, $\chi'(v, u)$ or $\frac{dx}{du}$ is to be found from equations (1) by eliminating y, regarding v as constant; the following is exactly equivalent; from (1)

$$\frac{d\phi_1}{dx}\frac{dx}{du}+\frac{d\phi_1}{dy}\frac{dy}{du}+\frac{d\phi_1}{du}=0,$$

$$\frac{d\phi_2}{dx}\frac{dx}{du}+\frac{d\phi_2}{dy}\frac{dy}{du}+\frac{d\phi_2}{du}=0.$$

From these equations by eliminating $\frac{dy}{du}$ we find

$$\frac{dx}{du} = \frac{\frac{d\phi_1}{du}\frac{d\phi_2}{dy} - \frac{d\phi_1}{dy}\frac{d\phi_2}{du}}{\frac{d\phi_1}{dy}\frac{d\phi_2}{dx} - \frac{d\phi_1}{dx}\frac{d\phi_2}{dy}}.$$

This then is an equivalent for $\chi'(v, u)$.

Thus $$\psi'(x, v)\,\chi'(v, u) = \frac{\frac{d\phi_1}{dv}\frac{d\phi_2}{du} - \frac{d\phi_1}{du}\frac{d\phi_2}{dv}}{\frac{d\phi_1}{dy}\frac{d\phi_2}{dx} - \frac{d\phi_1}{dx}\frac{d\phi_2}{dy}}.$$

Hence the conclusion is that

$$\iint V\,dx\,dy = \iint V \frac{\frac{d\phi_1}{dv}\frac{d\phi_2}{du} - \frac{d\phi_1}{du}\frac{d\phi_2}{dv}}{\frac{d\phi_1}{dy}\frac{d\phi_2}{dx} - \frac{d\phi_1}{dx}\frac{d\phi_2}{dy}}\,dv\,du \quad\ldots\ldots\ldots\ldots (4),$$

where after the differentiations have been performed, we must put for x and y their values in terms of u and v to be found from (1); also the values of x and y must be substituted in V.

An important particular case is that in which x and y are given *explicitly* as functions of u and v; the equations (1) then take the form

$$x - f_1(u, v) = 0, \qquad y - f_2(u, v) = 0 \quad\ldots\ldots\ldots (5).$$

Here $$\frac{d\phi_1}{dx} = 1, \quad \frac{d\phi_1}{dy} = 0, \quad \frac{d\phi_2}{dx} = 0, \quad \frac{d\phi_2}{dy} = 1,$$

and the transformed integral becomes

$$\iint V\left(\frac{df_1}{du}\frac{df_2}{dv} - \frac{df_1}{dv}\frac{df_2}{du}\right) dv\,du,$$

where we must substitute for x and y their values from (5) in V.

Thus we may write

$$\iint V\,dx\,dy = \iint V\left(\frac{dx}{du}\frac{dy}{dv} - \frac{dx}{dv}\frac{dy}{du}\right) dv\,du \;\;\ldots\ldots\ldots(6).$$

The formulæ in (4) and (6) are those which are usually given; they contain a simple solution of the proposed problem in those cases where the limits of the new integrations are obvious. But in some examples the difficulty of determining the limits of the new integrations would be very great, and to ensure a correct result it would be necessary instead of using these formulæ, to carry on the process precisely in the manner indicated in the theory, by removing one of the old variables at a time.

240. The following is an example.

Required to transform $\int_0^a \int_0^b V\,dx\,dy$, having given

$$y + x = u, \qquad y = uv.$$

From the given equations we have

$$x = u(1 - v), \qquad y = uv;$$

thus $\dfrac{dx}{du} = 1 - v, \quad \dfrac{dx}{dv} = -u, \quad \dfrac{dy}{du} = v, \quad \dfrac{dy}{dv} = u;$

therefore $\dfrac{dx}{du}\dfrac{dy}{dv} - \dfrac{dx}{dv}\dfrac{dy}{du} = u(1 - v) + uv = u.$

Hence by equation (6) of Art. 239, we have

$$\int_0^a \int_0^b V\,dx\,dy = \iint Vu\,dv\,du;$$

but we have not determined the limits of the integrations with respect to u and v, so that the result is of little value. We will now solve this example by following the steps indicated in the theory given above.

From the given equations connecting the old and new variables we eliminate u; thus we have

$$y = \frac{vx}{1-v}, \quad \text{therefore } \frac{dy}{dv} = \frac{x}{(1-v)^2};$$

to the limits $y=0$ and $y=b$, correspond respectively $v=0$ and $v=\frac{b}{b+x}$; thus

$$\int_0^a \int_0^b V\,dx\,dy = \int_0^a \int_0^{\frac{b}{b+x}} V_1\,x\,(1-v)^{-2}\,dx\,dv.$$

We have now to change the *order* of integration in

$$\int_0^a \int_0^{\frac{b}{b+x}} V_1\,x\,(1-v)^{-2}\,dx\,dv.$$

This question has been solved in Art. 235; hence we obtain

$$\int_0^a \int_0^b V\,dx\,dy = \int_0^a \int_0^{\frac{b}{b+x}} V_1\,x\,(1-v)^{-2}\,dx\,dv$$

$$= \int_0^{\frac{b}{b+a}} \int_0^a V_1\,x\,(1-v)^{-2}\,dv\,dx + \int_{\frac{b}{b+a}}^1 \int_0^{\frac{b(1-v)}{v}} V_1\,x\,(1-v)^{-2}\,dv\,dx.$$

We have now to change x for u where

$$x = u\,(1-v), \qquad \frac{dx}{du} = 1-v;$$

thus we obtain

$$\int_0^{\frac{b}{b+a}} \int_0^{\frac{a}{1-v}} V'u\,dv\,du + \int_{\frac{b}{a+b}}^1 \int_0^{\frac{b}{v}} V'u\,dv\,du,$$

since to the limits 0 and a for x correspond respectively 0 and $\frac{a}{1-v}$ for u, and to the limits 0 and $\frac{b(1-v)}{v}$ for x correspond respectively 0 and $\frac{b}{v}$ for u.

If $a=b$ the transformed integral becomes

$$\int_0^{\frac{1}{2}}\int_0^{\frac{a}{1-v}} V'u\,dv\,du+\int_{\frac{1}{2}}^1\int_0^{\frac{a}{v}} V'u\,dv\,du.$$

If a is made infinite, these two terms combine into the single expression

$$\int_0^1\int_0^{\infty} V'u\,dv\,du.$$

241. *Second Example.* Required to transform

$$\int_0^c\int_0^{c-x} V\,dx\,dy,$$

having given $y+x=u$, $y=uv$.

Perform the whole operation as before; so that we put

$$y=\frac{vx}{1-v} \text{ and } \frac{dy}{dv}=\frac{x}{(1-v)^2}.$$

When $y=0$ we have $v=0$, and when $y=c-x$ we have $v=\dfrac{c-x}{c}$. Thus the integral is transformed into

$$\int_0^c\int_0^{\frac{c-x}{c}} V_1 x(1-v)^{-2}\,dx\,dv.$$

Now change the order of integration; thus we obtain

$$\int_0^1\int_0^{c(1-v)} V_1 x(1-v)^{-2}\,dv\,dx.$$

Now put $x=u(1-v)$ and $\dfrac{dx}{du}=1-v$; the limits of u will be 0 and c. Hence we have finally for the transformed integral

$$\int_0^1\int_0^c V'u\,dv\,du.$$

242. *Third Example.* Transform $\iint V\,dx\,dy$ to a double integral with the variables r and θ, supposing

$$x = r\cos\theta, \qquad y = r\sin\theta.$$

We may put θ for v and r for u in the general formulæ; thus

$$\frac{dx}{du}\frac{dy}{dv} - \frac{dx}{dv}\frac{dy}{du} = r\cos^2\theta + r\sin^2\theta = r;$$

and the transformed integral is

$$\iint V' r\,d\theta\,dr.$$

This is a transformation with which the student is probably already familiar; the limits must of course be so taken that every element which enters into the original integral shall also occur in the transformed integral.

A particular case of this example may be noticed. Suppose the integral to be

$$\iint \phi\,(ax + by)\,dx\,dy;$$

by the present transformation this becomes

$$\iint \phi\,\{kr\cos(\theta - \alpha)\}\,r d\theta\,dr,$$

where $k\cos\alpha = a$ and $k\sin\alpha = b$. Now put $\theta - \alpha = \theta'$, so that the integral becomes

$$\iint \phi\,(kr\cos\theta')\,r d\theta'\,dr;$$

then suppose $r\cos\theta' = x'$ and $r\sin\theta' = y'$ and the integral may be again changed to

$$\iint \phi\,(kx')\,dx'\,dy'.$$

Thus suppressing the accents we may write

$$\iint \phi\,(ax + by)\, dx\, dy = \iint \phi\,(kx)\, dx\, dy,$$

where $k = \sqrt{(a^2 + b^2)}$. The limits will generally be different in the two integrals; those on the right-hand side must be determined by special examination, corresponding to given limits on the left-hand side.

243. *Fourth Example.* Transform $\int_0^c \int_0^x V\, dx\, dy$, having given

$$x = au + bv, \qquad y = bu + av, \qquad a > b.$$

Eliminate u, thus $ay - bx = (a^2 - b^2)\, v$, and the first transformation gives

$$\frac{a^2 - b^2}{a} \int_0^c \int_{-\frac{bx}{a^2 - b^2}}^{\frac{x}{a+b}} V_1\, dx\, dv,$$

where V_1 is what V becomes when we put $\frac{bx}{a} + \frac{a^2 - b^2}{a} v$ for y. Next change the order of integration; this gives

$$\frac{a^2 - b^2}{a} \int_0^{\frac{c}{a+b}} \int_{(a+b)\,v}^{c} V_1\, dv\, dx + \frac{a^2 - b^2}{a} \int_{-\frac{bc}{a^2 - b^2}}^{0} \int_{-\frac{a^2 - b^2}{b} v}^{c} V_1\, dv\, dx.$$

We have now to change from x to u by means of the equation $x = au + bv$, which gives $\frac{dx}{du} = a$; the limits of u corresponding to the known limits of x are easily ascertained.

Thus we have finally for the transformed integral

$$(a^2 - b^2) \int_0^{\frac{c}{a+b}} \int_v^{\frac{c - bv}{a}} V'\, dv\, du + (a^2 - b^2) \int_{-\frac{bc}{a^2 - b^2}}^{0} \int_{-\frac{av}{b}}^{\frac{c - bv}{a}} V'\, dv\, du.$$

The correctness of the transformation may be verified by supposing V to be some simple function of x and y; for example, if V be unity, the value of the original or of the transformed integral is $\frac{c^2}{2}$.

244. *Fifth Example.* The area of a surface is given by the integral

$$\iint dx\, dy \sqrt{\left\{1+\left(\frac{dz}{dx}\right)^2+\left(\frac{dz}{dy}\right)^2\right\}} \qquad \text{(Art. 170)};$$

required to transform it into an integral with respect to θ and ϕ, having given

$$z = r\cos\theta, \qquad x = r\sin\theta\cos\phi, \qquad y = r\sin\theta\sin\phi.$$

From the known equation to the surface z is given in terms of x and y; hence by substituting we have an equation which gives r in terms of θ and ϕ.

We will first find the transformation for $dx\, dy$:

$$\frac{dx}{d\theta} = \frac{dr}{d\theta}\sin\theta\cos\phi + r\cos\theta\cos\phi,$$

$$\frac{dx}{d\phi} = \frac{dr}{d\phi}\sin\theta\cos\phi - r\sin\theta\sin\phi,$$

$$\frac{dy}{d\theta} = \frac{dr}{d\theta}\sin\theta\sin\phi + r\cos\theta\sin\phi,$$

$$\frac{dy}{d\phi} = \frac{dr}{d\phi}\sin\theta\sin\phi + r\sin\theta\cos\phi.$$

Hence $\frac{dx}{d\theta}\frac{dy}{d\phi} - \frac{dx}{d\phi}\frac{dy}{d\theta} = r\sin\theta\left(r\cos\theta + \frac{dr}{d\theta}\sin\theta\right)$;

thus $dx\, dy$ will be replaced by

$$r\sin\theta\left(r\cos\theta + \frac{dr}{d\theta}\sin\theta\right) d\phi\, d\theta.$$

We have next to transform

$$\sqrt{\left\{1+\left(\frac{dz}{dx}\right)^2+\left(\frac{dz}{dy}\right)^2\right\}}.$$

We have
$$\frac{dz}{d\theta} = \frac{dz}{dx}\frac{dx}{d\theta} + \frac{dz}{dy}\frac{dy}{d\theta},$$
$$\frac{dz}{d\phi} = \frac{dz}{dx}\frac{dx}{d\phi} + \frac{dz}{dy}\frac{dy}{d\phi}.$$

Also
$$\frac{dz}{d\theta} = \frac{dr}{d\theta}\cos\theta - r\sin\theta,$$
$$\frac{dz}{d\phi} = \frac{dr}{d\phi}\cos\theta.$$

Thus $\frac{dz}{dx}$ is a fraction of which the numerator is
$$\frac{dz}{d\theta}\frac{dy}{d\phi} - \frac{dz}{d\phi}\frac{dy}{d\theta},$$

that is, $\left(\frac{dr}{d\theta}\cos\theta - r\sin\theta\right)\left(\frac{dr}{d\phi}\sin\theta\sin\phi + r\sin\theta\cos\phi\right)$
$$-\frac{dr}{d\phi}\cos\theta\left(\frac{dr}{d\theta}\sin\theta\sin\phi + r\cos\theta\sin\phi\right),$$

that is,
$$-r\sin\phi\frac{dr}{d\phi} + r\sin\theta\cos\theta\cos\phi\frac{dr}{d\theta} - r^2\sin^2\theta\cos\phi,$$

and the denominator is
$$\frac{dx}{d\theta}\frac{dy}{d\phi} - \frac{dx}{d\phi}\frac{dy}{d\theta},$$

the value of which was found before; thus
$$\frac{dz}{dx} = \frac{r\sin\theta\cos\theta\cos\phi\frac{dr}{d\theta} - r\sin\phi\frac{dr}{d\phi} - r^2\sin^2\theta\cos\phi}{r\sin\theta\left(r\cos\theta + \sin\theta\frac{dr}{d\theta}\right)}.$$

Similarly
$$\frac{dz}{dy} = \frac{r\cos\phi\frac{dr}{d\phi} + r\sin\theta\cos\theta\sin\phi\frac{dr}{d\theta} - r^2\sin^2\theta\sin\phi}{r\sin\theta\left(r\cos\theta + \sin\theta\frac{dr}{d\theta}\right)};$$

therefore

$$1+\left(\frac{dz}{dx}\right)^2+\left(\frac{dz}{dy}\right)^2=\frac{r^4\sin^2\theta+r^2\left(\frac{dr}{d\phi}\right)^2+r^2\sin^2\theta\left(\frac{dr}{d\theta}\right)^2}{r^2\sin^2\theta\left(r\cos\theta+\sin\theta\frac{dr}{d\theta}\right)^2},$$

and finally the transformed integral is

$$\iint\sqrt{\left\{r^2\sin^2\theta+\left(\frac{dr}{d\phi}\right)^2+\sin^2\theta\left(\frac{dr}{d\theta}\right)^2\right\}}\,r\,d\phi\,d\theta.$$

245. There will be no difficulty now in the transformation of a triple integral. Suppose that V is a function of x, y, z, and that $\iiint V\,dx\,dy\,dz$ is to be transformed into a triple integral with respect to three new variables u, v, w, which are connected with x, y, z by three equations. From the investigation of Art. 239, we may anticipate that the result will take its simplest form when the old variables are given explicitly in terms of the new. Suppose then

$$x=f_1(u, v, w),\quad y=f_2(u, v, w),\quad z=f_3(u, v, w)\ldots\ldots(1).$$

We first transform the integral with respect to z into an integral with respect to w. During the integration for z we regard x and y as constants; theoretically then we should from (1) express z as a function of x, y, and w, by eliminating u and v; we should then find the differential coefficient of z with respect to w regarding x and y as constants. But we may obtain the required result by differentiating equations (1) as they stand,

thus

$$\frac{df_1}{du}\frac{du}{dw}+\frac{df_1}{dv}\frac{dv}{dw}+\frac{df_1}{dw}=0,$$

$$\frac{df_2}{du}\frac{du}{dw}+\frac{df_2}{dv}\frac{dv}{dw}+\frac{df_2}{dw}=0,$$

$$\frac{df_3}{du}\frac{du}{dw}+\frac{df_3}{dv}\frac{dv}{dw}+\frac{df_3}{dw}=\frac{dz}{dw}.$$

Eliminate $\frac{du}{dw}$ and $\frac{dv}{dw}$; thus we find

$$\frac{dz}{dw} = \frac{N}{\frac{df_1}{du}\frac{df_2}{dv} - \frac{df_2}{du}\frac{df_1}{dv}},$$

where $$N = \frac{df_3}{dw}\left(\frac{df_1}{du}\frac{df_2}{dv} - \frac{df_2}{du}\frac{df_1}{dv}\right) + \frac{df_1}{dw}\left(\frac{df_2}{du}\frac{df_3}{dv} - \frac{df_3}{du}\frac{df_2}{dv}\right) + \frac{df_2}{dw}\left(\frac{df_3}{du}\frac{df_1}{dv} - \frac{df_1}{du}\frac{df_3}{dv}\right).$$

Hence the integral is transformed into

$$\iiint V_1 \frac{N}{\frac{df_1}{du}\frac{df_2}{dv} - \frac{df_2}{du}\frac{df_1}{dv}}\, dx\, dy\, dw,$$

where V_1 indicates what V becomes when for z its value in terms of x, y and w is substituted. We must also determine the limits of w from the known limits of z. Next we may change the order of integration for y and w, and then proceed as before to remove y and introduce v. Then again we should change the order of integration for w and x and then for v and x, and finally remove x and introduce u. And in examples it might be advisable to go through the process step by step, in order to obtain the limits of the transformed integral.

We may however more simply ascertain the final formula thus. Transform the integral with respect to z into an integral with respect to w as above; then twice change the order of integration, so that we have

$$\iiint V_1 \frac{N}{\frac{df_1}{du}\frac{df_2}{dv} - \frac{df_2}{du}\frac{df_1}{dv}}\, dw\, dx\, dy.$$

Now we have to transform the double integral with respect to x and y into a double integral with respect to u and v by means of the first two of equations (1). Hence we know by Art. 239, that the symbol $dx\, dy$ will be replaced by

$$\left(\frac{df_1}{du}\frac{df_2}{dv} - \frac{df_2}{du}\frac{df_1}{dv}\right) dv\, du;$$

and the integral is finally transformed into

$$\iiint V'N\,dw\,dv\,du,$$

where V' is what V becomes when for x, y, and z, their values in terms of u, v, and w are substituted.

The student will now have no difficulty in investigating the more complex case, in which the old and new variables are connected by equations of the form

$$\left.\begin{aligned}\phi_1(x, y, z, u, v, w) = 0\\ \phi_2(x, y, z, u, v, w) = 0\\ \phi_3(x, y, z, u, v, w) = 0\end{aligned}\right\} \dots\dots\dots\dots\dots (2).$$

Here it will be found that

$$\frac{dz}{dw} = \frac{N_1}{D_1}, \quad \frac{dy}{dv} = \frac{N_2}{D_2}, \quad \frac{dx}{du} = \frac{N_3}{D_3};$$

also that $\qquad N_2 = D_1$, and $N_3 = D_2$.

Thus $\qquad \iiint V\,dx\,dy\,dz = \iiint V'\frac{N_1}{D_3}\,du\,dv\,dw$, where

$$N_1 = \frac{d\phi_1}{dw}\left(\frac{d\phi_2}{du}\frac{d\phi_3}{dv} - \frac{d\phi_3}{du}\frac{d\phi_2}{dv}\right) + \frac{d\phi_2}{dw}\left(\frac{d\phi_3}{du}\frac{d\phi_1}{dv} - \frac{d\phi_1}{du}\frac{d\phi_3}{dv}\right)$$
$$+ \frac{d\phi_3}{dw}\left(\frac{d\phi_1}{du}\frac{d\phi_2}{dv} - \frac{d\phi_2}{du}\frac{d\phi_1}{dv}\right),$$

and $-D_3$ is equal to a similar expression with x, y, z instead of u, v, w respectively.

It may happen that equations (2) will impose some restriction as to the way in which the transformations are to be effected. For example suppose we have

$$x + y + z - u = 0, \quad x + y - uv = 0, \quad y - uvw = 0.$$

From these equations we cannot express z in terms of w and x and y, and therefore we cannot begin by transforming from z to w. We may however begin by transforming from z to u or from z to v; or we may begin by transforming from x or y to u or v or w.

246. It may be instructive to illustrate these transformations geometrically. We begin with the double integral.

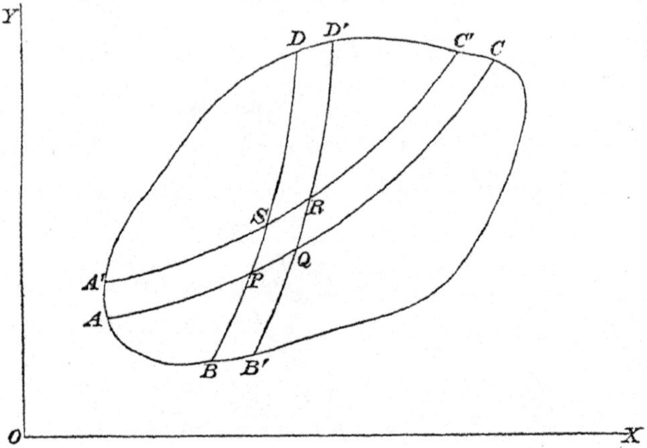

Let $\iint V\,dx\,dy$ be a double integral, which is to be taken for all the values of x and y comprised within the boundary $ABCD$. Suppose the variables x and y connected with two new variables u and v by the equations

$$x = f_1(u, v), \qquad y = f_2(u, v) \ldots\ldots\ldots\ldots\ldots(1).$$

From these equations let u and v be found in terms of x and y, so that we may write

$$u = F_1(x, y), \qquad v = F_2(x, y) \ldots\ldots\ldots\ldots\ldots(2).$$

Now by ascribing any constant value to u the first equation of (2) may be considered as representing a curve, and by giving in succession different constant values to u, we have a series of such curves. Let then $APQC$ be a curve, at every point of which $F_1(x, y)$ has a certain constant value u; and let $A'SRC'$ be a curve, at every point of which $F_1(x, y)$ has a certain constant value $u + \delta u$. Similarly let $BPSD$ be a curve, at every point of which $F_2(x, y)$ has a certain constant value v; and let $B'QRD'$ be a curve, at every point of which

$F_2(x, y)$ has a certain constant value $v + \delta v$. Let x, y now denote the co-ordinates of P; we shall proceed to express the co-ordinates of Q, S, and R.

The co-ordinates of Q are found from those of P, by changing v into $v + \delta v$; hence by (1) they are ultimately, when δv is indefinitely small,

$$x + \frac{dx}{dv}\delta v, \quad \text{and } y + \frac{dy}{dv}\delta v.$$

Similarly the co-ordinates of S are found from those of P by changing u into $u + \delta u$; hence by (1) they are ultimately

$$x + \frac{dx}{du}\delta u, \quad \text{and } y + \frac{dy}{du}\delta u.$$

The co-ordinates of R are found from those of P by changing both u into $u + \delta u$ and v into $v + \delta v$; hence by (1) they are ultimately

$$x + \frac{dx}{du}\delta u + \frac{dx}{dv}\delta v, \quad \text{and } y + \frac{dy}{du}\delta u + \frac{dy}{dv}\delta v.$$

These results shew that P, Q, R, S are ultimately situated at the angular points of a parallelogram. The area of this parallelogram may be taken without error in the limit for the area of the curvilinear figure $PQRS$. The expression for the area of the triangle PQR in terms of the co-ordinates of its angular points is known (see *Plane Co-ordinate Geometry*, Art. 11), and the area of the parallelogram is double that of the triangle. Hence we have ultimately for the area of $PQRS$ the expression

$$\pm\left(\frac{dx}{du}\frac{dy}{dv} - \frac{dx}{dv}\frac{dy}{du}\right)\delta u\,\delta v.$$

Thus it is obvious that the integral $\iint V\,dx\,dy$ may be replaced by

$$\pm\iint V'\left(\frac{dx}{du}\frac{dy}{dv} - \frac{dx}{dv}\frac{dy}{du}\right)du\,dv;$$

the ambiguity of sign would disappear in an example in which the limits of integration were known. In finding the value of the transformed integral, we may suppose that we first integrate with respect to v, so that u is kept constant; this amounts to taking all the elements such as $PQRS$, which form a strip such as $AA'C'C$. Then the integration with respect to u amounts to taking all such strips as $AA'C'C$ which are contained within the assigned boundary $ABCD$.

247. We proceed to illustrate geometrically the transformation of a triple integral.

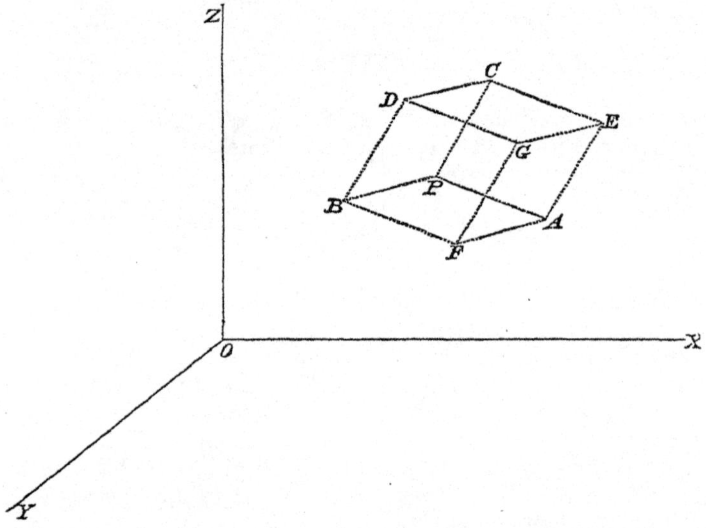

Let $\iiint V dx\,dy\,dz$ be a triple integral, which is to be taken for all values of x, y, and z comprised between certain assigned limits. Suppose the variables x, y, and z connected with three new variables u, v, w by the equations

$$x = f_1(u, v, w), \quad y = f_2(u, v, w), \quad z = f_3(u, v, w) \ldots\ldots(1).$$

From these equations let u, v, and w be found in terms of x, y, and z, so that we may write

$$u = F_1(x, y, z), \quad v = F_2(x, y, z), \quad w = F_3(x, y, z) \ldots.(2).$$

Now by ascribing any constant value to u, the first equation of (2) may be considered as representing a surface, and by giving in succession different constant values to u we have a series of such surfaces. Suppose there to be a surface at every point of which $F_1(x, y, z)$ has the constant value u, and let the four points P, B, D, C be in that surface; also suppose there to be a surface at every point of which $F_1(x, y, z)$ has the constant value $u+\delta u$, and let the four points A, F, G, E be in that surface. Similarly suppose P, A, E, C to be in a surface at every point of which $F_2(x, y, z)$ has the constant value v, and B, D, G, F to be in a surface at every point of which $F_2(x, y, z)$ has the constant value $v+\delta v$. Lastly suppose P, A, F, B to be in a surface at every point of which $F_3(x, y, z)$ has the constant value w, and C, D, G, E to be in a surface at every point of which $F_3(x, y, z)$ has the constant value $w+\delta w$.

Let x, y, z now denote the co-ordinates of P; we shall proceed to express the co-ordinates of the other points. The co-ordinates of A are found from those of P by changing u into $u+\delta u$; hence by (1) they are ultimately when δu is indefinitely small,

$$x+\frac{dx}{du}\delta u, \quad y+\frac{dy}{du}\delta u, \quad z+\frac{dz}{du}\delta u.$$

The co-ordinates of B are found from those of P by changing v into $v+\delta v$; hence by (1) they are ultimately

$$x+\frac{dx}{dv}\delta v, \quad y+\frac{dy}{dv}\delta v, \quad z+\frac{dz}{dv}\delta v.$$

Similarly the co-ordinates of C are ultimately

$$x+\frac{dx}{dw}\delta w, \quad y+\frac{dy}{dw}\delta w, \quad z+\frac{dz}{dw}\delta w.$$

The co-ordinates of D are found from those of P by changing v into $v+\delta v$, and w into $w+\delta w$; hence by (1) they are ultimately

$$x+\frac{dx}{dv}\delta v+\frac{dx}{dw}\delta w, \quad y+\frac{dy}{dv}\delta v+\frac{dy}{dw}\delta w, \quad z+\frac{dz}{dv}\delta v+\frac{dz}{dw}\delta w.$$

Similarly the co-ordinates of E, F and G may be found.

These results shew that P, A, B, C, D, E, F, G are ultimately situated at the angular points of a parallelopiped; and the volume of this parallelopiped may be taken without error in the limit for the volume of the solid bounded by the six surfaces which we have referred to. Now by a known theorem the volume of a tetrahedron can be expressed in terms of the co-ordinates of its angular points, and the volume of the parallelopiped PG is six times that of the tetrahedron $ABPC$. Hence finally we have for the volume of the parallelopiped

$$\pm\left\{\frac{dx}{du}\left(\frac{dy}{dv}\frac{dz}{dw}-\frac{dy}{dw}\frac{dz}{dv}\right)+\frac{dy}{du}\left(\frac{dz}{dv}\frac{dx}{dw}-\frac{dz}{dw}\frac{dx}{dv}\right)\right.$$
$$\left.+\frac{dz}{du}\left(\frac{dx}{dv}\frac{dy}{dw}-\frac{dx}{dw}\frac{dy}{dv}\right)\right\}\delta u\,\delta v\,\delta w=\pm N\delta u\,\delta v\,\delta w \text{ say.}$$

Hence the triple integral is transformed into

$$\pm\iiint V' N du\,dv\,dw;$$

the ambiguity in sign would disappear in an example where the limits of integration were known.

248. We have now given the theory of the transformation of double and triple integrals; the essential point in our investigation is, that we have shewn how to remove the old variables and replace them by the new variables *one at a time*. We recommend the student to pay attention to this point, as we conceive that the theory of the subject is thus made clear and simple, and at the same time the limits of the transformed integral can be more easily ascertained. We do not lay any stress on the geometrical *illustrations* in the two preceding articles; they require much more development before they can be accepted as rigid *demonstrations*.

249. Before leaving the subject we will briefly indicate the method formerly used in solving the problem. This method we have not brought prominently forward, partly because it gives no assistance in determining the new limits,

and partly on account of its obscurity; the latter defect has been frequently noticed by writers on the subject.

Suppose $\iint V dx dy$ is to be transformed into an integral with respect to two new variables u and v of which the old variables are known functions.

Let the variables undergo infinitesimal changes: thus

$$dx = \frac{dx}{du} du + \frac{dx}{dv} dv \text{ .... (1)},$$

$$dy = \frac{dy}{du} du + \frac{dy}{dv} dv \text{ .... (2)}.$$

Now in the original expression $V dx\, dy$ in forming dx we suppose y *constant*, that is, $dy = 0$; hence (2) becomes

$$0 = \frac{dy}{du} du + \frac{dy}{dv} dv \text{ .... (3)},$$

find dv from this and substitute it in (1); thus

$$dx = \frac{\frac{dx}{du}\frac{dy}{dv} - \frac{dx}{dv}\frac{dy}{du}}{\frac{dy}{dv}} du \text{ .... (4)}.$$

Again, in forming dy in $V dx\, dy$ we suppose x constant, that is, $dx = 0$; hence by (4) we must suppose $du = 0$; thus from (2)

$$dy = \frac{dy}{dv} dv \text{ .... (5)}.$$

From (4) and (5)

$$dx\, dy = \left(\frac{dx}{du}\frac{dy}{dv} - \frac{dx}{dv}\frac{dy}{du}\right) du\, dv;$$

and $\iint V dx\, dy$ becomes

$$\iint V' \left(\frac{dx}{du}\frac{dy}{dv} - \frac{dx}{dv}\frac{dy}{du}\right) du\, dv.$$

With respect to the limits of integration we can onl give the general direction, that the new limits must be s taken as to include every element which was included by th old limits.

250. Similarly in transforming a triple integral

$$\iiint V\, dx\, dy\, dz$$

the process was as follows. Let the new variables be u, v, w in forming dz we must suppose x and y constant; thus w have

$$dz = \frac{dz}{du} du + \frac{dz}{dv} dv + \frac{dz}{dw} dw,$$

$$0 = \frac{dx}{du} du + \frac{dx}{dv} dv + \frac{dx}{dw} dw,$$

$$0 = \frac{dy}{du} du + \frac{dy}{dv} dv + \frac{dy}{dw} dw,$$

thus $$dz = \frac{N\, dw}{\frac{dx}{du}\frac{dy}{dv} - \frac{dx}{dv}\frac{dy}{du}} \quad \text{.......................} \quad (1),$$

where N has the same value as in Art. 247.

Next in forming dy we have to regard x and z as constant; hence by (1) we must regard w as constant; thus we have

$$dy = \frac{dy}{du} du + \frac{dy}{dv} dv,$$

$$0 = \frac{dx}{du} du + \frac{dx}{dv} dv;$$

therefore $$dy = \frac{\left(\frac{dy}{dv}\frac{dx}{du} - \frac{dy}{du}\frac{dx}{dv}\right) dv}{\frac{dx}{du}} \quad \text{.................} \quad (2).$$

And lastly in forming dx we suppose y and z constant, that is, by (1) and (2) we suppose w and v constant; thus

$$dx = \frac{dx}{du} du \qquad\text{..........................(3).}$$

From (1), (2), and (3)

$$dx\,dy\,dz = N du\,dv\,dw.$$

251. The student who wishes to study the history of this subject may be assisted by the following references. Lacroix, *Calcul Diff. et Intégral*, Vol. II. p. 205; also the references to the older authorities will be found in page XI. of the table prefixed to this volume. De Morgan, *Diff. and Integral Calculus*, p. 392. Moigno, *Calcul Diff. et Intégral*, Vol. II. p. 214; Ostrogradsky, *Mémoires de l'Académie de St Pétersbourg*, Sixième Série, 1838, p. 401. Catalan, *Mémoires Couronnés par l'Académie...de Bruxelles*, Vol. XIV. p. 1. Boole, *Cambridge Mathematical Journal*, Vol. IV. p. 20. Cauchy, *Exercices d'Analyse et de Physique Mathématique*, Vol. IV. p. 128. Svanberg, *Nova Acta Regiæ Societatis Scientiarum Upsaliensis*, Vol. XIII. 1847, p. 1. De Morgan, *Transactions of the Cambridge Phil. Society*, Vol. IX. p. [133].

EXAMPLES.

1. Shew that if $x = a \sin\theta \sin\phi$ and $y = b \cos\theta \sin\phi$, the double integral $\iint dx\,dy$ is transformed into

$$\pm \iint ab \sin\phi \cos\phi \, d\phi \, d\theta.$$

2. If $x = u \sin\alpha + v \cos\alpha$ and $y = u \cos\alpha - v \sin\alpha$, prove that

$$\iint f(x, y) \frac{dx\,dy}{\sqrt{(1 - x^2 - y^2)}} = \iint f_1 (u, v) \frac{du\,dv}{\sqrt{(1 - u^2 - v^2)}}.$$

3. Prove that

$$\int_0^\infty \int_0^\infty \phi\,(a^2x^2 + b^2y^2)\,dx\,dy = \frac{\pi}{4ab}\int_0^\infty \phi(x)\,dx.$$

4. Transform $\iint V\,dx\,dy$, where $y = xu$ and $x = \dfrac{v}{1+u}$.

If the limits of y be 0 and x and the limits of x be 0 and a, find the limits in the transformed integral.

Result. $\displaystyle\int_0^1 \int_0^{a(1+u)} V'v\,(1+u)^{-2}\,du\,dv.$

5. Transform $\iint e^{-(x^2+2xy\cos\alpha+y^2)}\,dx\,dy$ from rectangular to polar co-ordinates, and thence shew that if the limits both of x and y be zero and infinity, the value of the integral will be $\dfrac{\alpha}{2\sin\alpha}$.

6. Transform $\displaystyle\int_0^a \int_0^b \phi\,(x, y)\,dx\,dy$ to polar co-ordinates, and indicate the limits for each order in the transformed integral.

Shew that

$$\int_0^a \int_0^b \frac{dx\,dy}{(c^2 + x^2 + y^2)^{\frac{3}{2}}} = \frac{1}{c}\tan^{-1}\frac{ab}{c\sqrt{(a^2 + b^2 + c^2)}}.$$

7. Apply the transformation from rectangular to polar co-ordinates in double integrals to shew that

$$\int_{-\infty}^{+\infty} \int_{-\infty}^{+\infty} \frac{a\,dx\,dy}{(x^2+y^2+a^2)^{\frac{3}{2}}\,(x^2+y^2+a'^2)^{\frac{1}{2}}} = \frac{2\pi}{a+a'}.$$

8. Transform the double integral $\iint f(x, y)\,dx\,dy$ into one

in which r and θ shall be the independent variables, having given

$$x = r\cos\theta + a\sin\theta, \qquad y = r\sin\theta + a\cos\theta.$$

Result.

$$\iint f(r\cos\theta + a\sin\theta,\ r\sin\theta + a\cos\theta)\,(a\sin 2\theta - r)\,d\theta\,dr.$$

9. Transform $\iint e^{-x^2-y^2}dx\,dy$ into a double integral where r and t are the independent variables where $\frac{y}{x} = t$ and $r^2 = x^2 + y^2$; and if the limits of x and y be each 0 and ∞, find the limits of r and t.

Result. $\displaystyle\int_0^\infty\int_0^\infty \frac{e^{-r^2}r\,dr\,dt}{1+t^2}$.

10. If x and y are given as functions of r and θ, transform the integral $\iiint dx\,dy\,dz$ into another where r, θ and z are the variables; and if $x = r\cos\theta$ and $y = r\sin\theta$, find the volume included by the four surfaces whose equations are $r = a$, $z = 0$, $\theta = 0$, and $z = mr\cos\theta$.

Result. The volume $\displaystyle = \int_0^{\frac{\pi}{2}}\int_0^a r^2 m\cos\theta\,d\theta\,dr = \frac{ma^3}{3}$.

11. If $\alpha x = yz$, $\beta y = zx$, $\gamma z = xy$, shew that

$$\iiint f(\alpha, \beta, \gamma)\,d\alpha\,d\beta\,d\gamma = 4\iiint f\left(\frac{yz}{x}, \frac{zx}{y}, \frac{xy}{z}\right)dx\,dy\,dz.$$

12. Transform $\iiiint V\,dx_1\,dx_2\,dx_3\,dx_4$ to r, θ, ϕ and ψ when

$$x_1 = r\sin\theta\cos\phi, \qquad x_3 = r\cos\theta\cos\psi,$$
$$x_2 = r\sin\theta\sin\phi, \qquad x_4 = r\cos\theta\sin\psi.$$

Result. $\displaystyle\iiiint V' r^3\sin\theta\cos\theta\,dr\,d\theta\,d\phi\,d\psi$.

13. Find the elementary area included between the curves $\phi(x, y) = u$, $\psi(x, y) = v$, and the curves obtained by giving to the parameters u and v indefinitely small increments.

Find the area included between a parabola and the tangents at the extremities of the latus rectum by dividing the area by a series of parabolas which touch these tangents and by a series of lines drawn from the intersection of the tangents.

14. Transform the triple integral $\iiint f(x, y, z)\, dx\, dy\, dz$ into one in which r, y, z are the independent variables, having given $\psi(x, y, z, r) = 0$; and change the variables in the above integral from x, y, z to r, θ, ϕ, having given

$$\psi(x, y, z, r) = 0, \quad \psi_1(y, z, r, \theta) = 0, \quad \psi_2(z, r, \theta, \phi) = 0.$$

Result. $$-\iiint \frac{\dfrac{d\psi}{dr}\dfrac{d\psi_1}{d\theta}\dfrac{d\psi_2}{d\phi}}{\dfrac{d\psi}{dx}\dfrac{d\psi_1}{dy}\dfrac{d\psi_2}{dz}} f_1(r, \theta, \phi)\, dr\, d\theta\, d\phi.$$

15. Transform the double integral

$$\iint dx\, dy \sqrt{\left\{1 + \left(\frac{dz}{dx}\right)^2 + \left(\frac{dz}{dy}\right)^2\right\}},$$

in which x, y, z are connected by the equation $x^2 + y^2 + z^2 = 1$, to an integral in terms of θ and ϕ, having these relations,

$$x = \sin\phi \sqrt{(1 - m^2 \sin^2\theta)}, \qquad y = \cos\theta \cos\phi,$$

$$z = \sin\theta \sqrt{(1 - n^2 \sin^2\phi)}, \qquad m^2 + n^2 = 1.$$

Hence prove that

$$\int_0^{\frac{\pi}{2}}\int_0^{\frac{\pi}{2}} \frac{m^2 \cos^2\theta + n^2 \cos^2\phi}{\sqrt{(1 - m^2 \sin^2\theta)}\sqrt{(1 - n^2 \sin^2\phi)}}\, d\theta\, d\phi = \frac{\pi}{2}.$$

16. Transform the integral $\iiint dx\,dy\,dz$ to r, θ, ϕ, where

$$x = r\sin\phi\sqrt{(1-n^2\cos^2\theta)}, \qquad y = r\cos\phi\sin\theta,$$
$$z = r\cos\theta\sqrt{(\cos^2\phi + n^2\sin^2\phi)}.$$

Result. $$\iiint \frac{r^2\{(n^2-1)\cos^2\phi - n^2\sin^2\theta\}\,dr\,d\theta\,d\phi}{\sqrt{(1-n^2\cos^2\theta)}\sqrt{(\cos^2\phi+n^2\sin^2\phi)}}.$$

17. Transform the expression $\iint \frac{r^3}{3}\sin\theta\,d\theta\,d\phi$ for a volume, to rectangular co-ordinates.

Result. $\frac{1}{3}\iint (z - px - qy)\,dx\,dy$; this should be interpreted geometrically.

18. If $x+y+z=u$, $x+y=uv$, $y=uvw$, prove that

$$\int_0^\infty\int_0^\infty\int_0^\infty V\,dx\,dy\,dz = \int_0^\infty\int_0^1\int_0^1 Vu^2v\,du\,dv\,dw.$$

19. If
$$x_1 = r\cos\theta_1,$$
$$x_2 = r\sin\theta_1\cos\theta_2,$$
$$x_3 = r\sin\theta_1\sin\theta_2\cos\theta_3,$$
$$\dots\dots\dots\dots\dots\dots$$
$$x_{n-1} = r\sin\theta_1\sin\theta_2\dots\sin\theta_{n-2}\cos\theta_{n-1},$$
$$x_n = r\sin\theta_1\sin\theta_2\dots\sin\theta_{n-2}\sin\theta_{n-1},$$

shew that $\iiint \dots\dots V\,dx_1\,dx_2\dots dx_n$

$$= \pm\iiint\dots\dots V' r^{n-1}(\sin\theta_1)^{n-2}(\sin\theta_2)^{n-3}$$
$$\dots\dots\sin\theta_{n-2}\,dr\,d\theta_1\,d\theta_2\dots\dots d\theta_{n-1},$$

where V is any function of $x_1, x_2, \dots x_n$, and V' what this function becomes when the variables are changed.

CHAPTER XII.

DEFINITE INTEGRALS.

252. WHEN the indefinite integral of a function is known, we can immediately obtain the value of the *definite integral* corresponding to any assigned limits of the variable. Sometimes however we are able by special methods to assign the value of a *definite* integral when we cannot express the indefinite integral in a finite form; sometimes without actually finding the value of a definite integral we can shew that it possesses important properties. In some cases in which the indefinite integral of a function can be found, the definite integral between certain limits may have a value which is worthy of notice, on account of the simple form in which it may be expressed. We shall in the present chapter give examples of these general statements.

We may observe that a collection of the known results with respect to Definite Integrals has been published in a quarto volume at Amsterdam, by D. Bierens de Haan, under the title of *Tables d'Intégrales Définies*.

253. Suppose $f(x)$ and $F(x)$ rational algebraical functions of x, and $f(x)$ of lower dimensions than $F(x)$, and suppose the equation $F(x) = 0$ to have no real roots; it is required to find the value of

$$\int_{-\infty}^{\infty} \frac{f(x)}{F(x)}\, dx.$$

It will be seen that under the above suppositions, the expression to be integrated never becomes infinite for real values of x.

Let $\alpha+\beta\sqrt{(-1)}$ and $\alpha-\beta\sqrt{(-1)}$ represent a pair of the imaginary roots of $F(x)=0$; then the corresponding quadratic fraction of the series into which $\frac{f(x)}{F(x)}$ can be decomposed, may be represented by

$$\frac{2A(x-\alpha)+2B\beta}{(x-\alpha)^2+\beta^2},$$

the constants A and B being found from the equation

$$A-B\sqrt{(-1)}=\frac{f\{\alpha+\beta\sqrt{(-1)}\}}{F'\{\alpha+\beta\sqrt{(-1)}\}} \qquad \text{(Art. 21).}$$

Now
$$\int\frac{2B\beta\,dx}{(x-\alpha)^2+\beta^2}=2B\tan^{-1}\frac{x-\alpha}{\beta},$$

therefore
$$\int_{-\infty}^{\infty}\frac{2B\beta\,dx}{(x-\alpha)^2+\beta^2}=2B\pi.$$

Also
$$\int_{-\infty}^{\infty}\frac{(x-\alpha)\,dx}{(x-\alpha)^2+\beta^2}=\int_{-\infty}^{\infty}\frac{tdt}{t^2+\beta^2},$$

and it is obvious that the latter integral between the assigned limits is zero, for the negative part is numerically equal to the positive part. Thus $2B\pi$ represents the part of the integral corresponding to the pair of imaginary roots under consideration.

If then we suppose $F(x)$ to be of $2n$ dimensions, and $B_1, B_2, \ldots\ldots B_n$ to be the n terms of which we have taken B as the type, we have

$$\int_{-\infty}^{\infty}\frac{f(x)}{F(x)}\,dx=2\pi\{B_1+B_2+\ldots\ldots+B_n\}.$$

254. As an example of the preceding article we take

$$\int_{-\infty}^{\infty}\frac{x^{2m}\,dx}{1+x^{2n}},$$

where m and n are positive integers, and m less than n. Here

$$A-B\sqrt{(-1)}=\frac{1}{2n\{\alpha+\beta\sqrt{(-1)}\}^{2n-2m-1}},$$

and it is known from the theory of equations that the values of $\alpha+\beta\sqrt{(-1)}$ are obtained from the expression

$$\cos\frac{(2r+1)\pi}{2n}+\sqrt{(-1)}\sin\frac{(2r+1)\pi}{2n},$$

by giving to r successively the values 0, 1, 2,...... up to $n-1$.

Thus, by Demoivre's theorem,

$$\{\alpha+\beta\sqrt{(-1)}\}^{2n-2m-1}=\cos\phi+\sqrt{(-1)}\sin\phi,$$

where

$$\phi=(2n-2m-1)\frac{(2r+1)\pi}{2n}=(2r+1)\pi-(2r+1)\frac{(2m+1)\pi}{2n};$$

so that

$$\cos\phi+\sqrt{(-1)}\sin\phi=-\cos(2r+1)\theta+\sqrt{(-1)}\sin(2r+1)\theta,$$

where $$\theta=\frac{2m+1}{2n}\pi.$$

Hence

$$A-B\sqrt{(-1)}=\frac{1}{2n}\frac{1}{-\cos(2r+1)\theta+\sqrt{(-1)}\sin(2r+1)\theta}$$

$$=-\frac{\cos(2r+1)\theta+\sqrt{(-1)}\sin(2r+1)\theta}{2n};$$

therefore $$B=\frac{\sin(2r+1)\theta}{2n}.$$

Hence

$$\int_{-\infty}^{\infty}\frac{x^{2m}\,dx}{1+x^{2n}}=\frac{\pi}{n}\{\sin\theta+\sin 3\theta+\sin 5\theta+\ldots+\sin(2n-1)\theta\}.$$

The sum of the series of sines is shewn in works on Trigonometry to be $\frac{\sin^2 n\theta}{\sin\theta}$, and in the present case $n\theta=\frac{2m+1}{2}\pi$, so that $\sin^2 n\theta=1$. Therefore

$$\int_{-\infty}^{\infty}\frac{x^{2m}\,dx}{1+x^{2n}}=\frac{\pi}{n\sin\frac{2m+1}{2n}\pi}.$$

It is obvious that $\int_0^\infty \frac{x^{2m}\,dx}{1+x^{2n}}$ is half of the above result, that is,

$$\int_0^\infty \frac{x^{2m}\,dx}{1+x^{2n}} = \frac{\pi}{2n \sin \frac{2m+1}{2n} \pi}.$$

255. In the last formula of the preceding article put $x^{2n} = y$, and suppose $\frac{2m+1}{2n} = k$; thus we obtain

$$\int_0^\infty \frac{y^{k-1}\,dy}{1+y} = \frac{\pi}{\sin k\pi} \quad \text{.................(1)}.$$

This result holds when k has any value comprised between 0 and 1. For the only restriction on the positive integers m and n is that m must be less than n, and therefore by properly choosing m and n we may make $\frac{2m+1}{2n}$ equal to any assigned proper fraction which has an *even* denominator when in its lowest terms. And although we cannot make $\frac{2m+1}{2n}$ *exactly* equal to any fraction which has an *odd* denominator when in its lowest terms, yet we can make it differ from such a fraction by as small a quantity as we please, and thus deduce the required result.

In the last result put x^r for y, where r is any positive quantity; thus

$$\int_0^\infty \frac{r x^{kr-r} x^{r-1}\,dx}{1+x^r} = \frac{\pi}{\sin k\pi},$$

that is,

$$\int_0^\infty \frac{x^{kr-1}\,dx}{1+x^r} = \frac{\pi}{r \sin k\pi}.$$

Let $kr = s$; thus

$$\int_0^\infty \frac{x^{s-1}\,dx}{1+x^r} = \frac{\pi}{r \sin \frac{s}{r} \pi}.$$

The only restriction on the positive quantities r and s is that s must be less than r.

The student will probably find no serious difficulty in the method we have indicated for proving the truth of equation (1) when k is a fraction which has an *odd* denominator when in its lowest terms; nevertheless a few remarks may be made which will establish the proposition decisively, and which will also serve as useful exercises in the subject of the present chapter.

Let $$u = \int_0^\infty \frac{y^{k-1}\,dy}{1+y}; \text{ then}$$

$$u = \int_0^1 \frac{y^{k-1}\,dy}{1+y} + \int_1^\infty \frac{y^{k-1}\,dy}{1+y};$$

and by putting $\frac{1}{z}$ for y we find that

$$\int_1^\infty \frac{y^{k-1}\,dy}{1+y} = \int_0^1 \frac{z^{-k}\,dz}{1+z}; \text{ thus}$$

$$u = \int_0^1 \frac{y^{k-1}+y^{-k}}{1+y}\,dy.$$

Therefore $$\frac{du}{dk} = \int_0^1 \frac{\log y}{1+y}(y^{k-1}-y^{-k})\,dy \text{..........(2)}.$$

Equation (2) shews that $\frac{du}{dk}$ is negative if $y^{k-1}-y^{-k}$ is constantly positive, and positive if $y^{k-1}-y^{-k}$ is constantly negative, between the limits 0 and 1 for y. Hence $\frac{du}{dk}$ is negative or positive according as k is less or greater than $\frac{1}{2}$. Thus u diminishes as k increases from 0 to $\frac{1}{2}$, and u increases as k increases from $\frac{1}{2}$ to 1.

Now let $\frac{\alpha}{\beta}$ denote any fraction in its lowest terms, in which β is an odd integer; and let p be any even integer. Let $k_1 = \frac{p\alpha - 1}{p\beta}$, and $k_3 = \frac{p\alpha + 1}{p\beta}$, and let k_2 denote $\frac{\alpha}{\beta}$. Let u_1, u_2, u_3 denote the values of $\int_0^\infty \frac{y^{k-1}\,dy}{1+y}$ when for k we substitute k_1, k_2, k_3 respectively. Then by equation (1)

$$u_1 = \frac{\pi}{\sin k_1\pi} \text{ and } u_3 = \frac{\pi}{\sin k_3\pi}.$$

Now we may take p so large that k_1 and k_3 shall be both greater or both less than $\frac{1}{2}$; and then by the inferences drawn from equation (2) it follows that u_2 must lie numerically between u_1 and u_3. Thus u_2 cannot differ from u_1 or u_3 by so much as the difference of u_1 and u_3; and therefore *a fortiori* u_2 cannot differ from $\frac{\pi}{\sin k_2\pi}$ by so much as the difference of u_1 and u_3. Hence as p may be indefinitely increased we have finally $u_2 = \frac{\pi}{\sin k_2\pi}$.

Eulerian Integrals.

256. The definite integral

$$\int_0^1 x^{l-1}(1-x)^{m-1}\,dx$$

is called the *first Eulerian integral;* we shall denote it by the symbol $B(l, m)$.

The definite integral

$$\int_0^\infty e^{-x} x^{n-1}\,dx$$

is called the *second Eulerian integral;* it is denoted by the symbol $\Gamma(n)$.

We shall now give some of the properties of these integrals; the constants in these integrals, which we have denoted by l, m, n, are supposed *positive* in all that follows.

257. In the first Eulerian integral put $x = 1 - z$;

thus $$\int_0^1 x^{l-1} (1-x)^{m-1} dx = \int_0^1 z^{m-1} (1-z)^{l-1} dz;$$

this shews that the constants l and m may be interchanged without altering the value of the integral; that is,

$$B(l, m) = B(m, l).$$

Again in the first Eulerian integral put $x = \frac{y}{1+y}$; thus

$$\int_0^1 x^{l-1} (1-x)^{m-1} dx = \int_0^\infty \frac{y^{l-1} dy}{(1+y)^{l+m}}.$$

In the same integral put $x = \frac{1}{1+y}$; thus

$$\int_0^1 x^{l-1} (1-x)^{m-1} dx = \int_0^\infty \frac{y^{m-1} dy}{(1+y)^{l+m}}.$$

258. Let $e^{-x} = y$, so that $x = \log \frac{1}{y}$; then we have

$$\int_0^\infty e^{-x} x^{n-1} dx = \int_0^1 \left(\log \frac{1}{y}\right)^{n-1} dy,$$

which consequently gives another form of $\Gamma(n)$.

259. We have by integration by parts

$$\int e^{-x} x^n dx = -e^{-x} x^n + n \int e^{-x} x^{n-1} dx;$$

and $e^{-x} x^n$ vanishes when $x = 0$, and also when $x = \infty$. (See *Diff. Cal.* Art. 153); thus

$$\int_0^\infty e^{-x} x^n dx = n \int_0^\infty e^{-x} x^{n-1} dx;$$

that is, $$\Gamma(n+1) = n\Gamma(n) \qquad (1).$$

Since $\int e^{-x}\,dx = -e^{-x}$ we have $\int_0^\infty e^{-x}\,dx = 1$; that is,

$$\Gamma(1) = 1 \qquad\qquad (2).$$

From (1) and (2) we see that if n be an integer

$$\Gamma(n+1) = \lfloor n.$$

When n is not an integer we may by repeated use of equation (1) make the value of $\Gamma(n)$ where n is greater than unity depend on that of $\Gamma(m)$ where m is less than unity.

260. By assuming $kx = z$ we have

$$\int_0^\infty e^{-kx} x^{n-1}\,dx = \frac{1}{k^n}\int_0^\infty e^{-z} z^{n-1}\,dz = \frac{\Gamma(n)}{k^n}.$$

261. We shall now prove an important equation which connects the two Eulerian integrals.

Integrate the double integral $\int_0^\infty\int_0^\infty x^{l+m-1} y^{m-1} e^{-(1+y)x}\,dy\,dx$ first with respect to x; we thus obtain, by Art. 260,

$$\Gamma(l+m)\int_0^\infty \frac{y^{m-1}\,dy}{(1+y)^{l+m}}.$$

Again integrate the same double integral first with respect to y; we thus obtain

$$\Gamma(m)\int_0^\infty \frac{e^{-x} x^{l+m-1}}{x^m}\,dx,$$

that is, $$\Gamma(m)\int_0^\infty e^{-x} x^{l-1}\,dx,$$

that is, $$\Gamma(m)\,\Gamma(l).$$

Hence $$\int_0^\infty \frac{y^{m-1}\,dy}{(1+y)^{l+m}} = \frac{\Gamma(l)\,\Gamma(m)}{\Gamma(l+m)}.$$

Hence, by Art. 257,

$$B(l, m) = \frac{\Gamma(l)\,\Gamma(m)}{\Gamma(l+m)}.$$

262. In the result of the preceding article, suppose $l+m=1$; thus, if m is less than unity,

$$\int_0^\infty \frac{y^{m-1}\,dy}{1+y} = \Gamma(m)\,\Gamma(1-m),$$

since $\Gamma(1)=1$. Hence, by Art. 255, if m is less than unity

$$\Gamma(m)\,\Gamma(1-m) = \frac{\pi}{\sin m\pi}.$$

263. Put $m=\frac{1}{2}$ in the last result; then

$$\Gamma(\tfrac{1}{2})\,\Gamma(\tfrac{1}{2}) = \pi,$$

therefore $$\Gamma(\tfrac{1}{2}) = \sqrt{\pi}.$$

We will give another proof of the last result.

Let $u = \int_0^\infty e^{-x^2}\,dx$; then it is obvious that u also

$$= \int_0^\infty e^{-y^2}\,dy;$$

thus $$u^2 = \int_0^\infty e^{-x^2}\,dx \times \int_0^\infty e^{-y^2}\,dy$$

$$= \int_0^\infty \int_0^\infty e^{-x^2-y^2}\,dx\,dy \qquad \text{(Art. 66)}.$$

This double integral is shewn in Art. 204 to be

$$= \tfrac{1}{4}\int_0^{2\pi}\int_0^\infty e^{-r^2}\,r\,d\theta\,dr = \frac{\pi}{4},$$

therefore $$u = \frac{\sqrt{\pi}}{2}.$$

Now $$\Gamma(\tfrac{1}{2}) = \int_0^\infty e^{-x}\,x^{-\frac{1}{2}}\,dx; \text{ put } x=y^2,$$

thus $$\Gamma(\tfrac{1}{2}) = 2\int_0^\infty e^{-y^2}\,dy = 2u = \sqrt{\pi}.$$

264. We shall now give an expression for $\Gamma(n)$ that will afford another proof of the result in Art. 262. We know that

the limit of $\frac{x^h - 1}{h}$ when h is indefinitely diminished is $\log x$; hence

$$\left(\log \frac{1}{x}\right)^{n-1} = \text{limit of } \left(\frac{1-x^h}{h}\right)^{n-1};$$

so we may write

$$\left(\log \frac{1}{x}\right)^{n-1} = \left(\frac{1-x^h}{h}\right)^{n-1} + y,$$

where y is a quantity that diminishes without limit when h does so.

Put $h = \frac{1}{r}$, then, by Art. 258,

$$\Gamma(n) = r^{n-1} \int_0^1 (1 - x^{\frac{1}{r}})^{n-1} dx + \int_0^1 y\, dx.$$

In the first integral put $x = z^r$; thus

$$\Gamma(n) - \int_0^1 y\, dx = r^n \int_0^1 z^{r-1} (1-z)^{n-1} dz.$$

We have it in our power to suppose r an integer; then the integral on the right-hand side, by Art. 33, is

$$\frac{1 \,.\, 2 \,.\, 3 \ldots\ldots r}{n(n+1)\ldots\ldots(n+r-1)} r^{n-1}.$$

Let r increase indefinitely, then y vanishes and we have

$$\Gamma(n) = \text{limit of } \frac{1 \,.\, 2 \,.\, 3 \ldots\ldots r}{n(n+1)\ldots\ldots(n+r-1)} r^{n-1}.$$

265. From the result of the preceding article we have

$$\frac{\{\Gamma(n)\}^2}{\Gamma(n-m)\,\Gamma(n+m)} = \left\{1 - \frac{m^2}{n^2}\right\}\left\{1 - \frac{m^2}{(n+1)^2}\right\}\left\{1 - \frac{m^2}{(n+2)^2}\right\}\ldots$$

A particular case of this is obtained by supposing $n = 1$; thus

$$\frac{1}{\Gamma(1-m)\,\Gamma(1+m)} = \left(1 - \frac{m^2}{1^2}\right)\left(1 - \frac{m^2}{2^2}\right)\left(1 - \frac{m^2}{3^2}\right)\ldots\ldots\ldots;$$

the expression on the right-hand side is known to be equal to $\frac{\sin m\pi}{m\pi}$; thus

$$\Gamma(1-m)\,\Gamma(1+m)=\frac{m\pi}{\sin m\pi},$$

therefore $$\Gamma(m)\,\Gamma(1-m)=\frac{\pi}{\sin m\pi} \qquad \text{(Art. 259).}$$

266. We shall now establish the following equation, n being an integer,

$$\Gamma\left(\frac{1}{n}\right)\Gamma\left(\frac{2}{n}\right)\Gamma\left(\frac{3}{n}\right)\ldots\ldots\Gamma\left(\frac{n-1}{n}\right)=(2\pi)^{\frac{n-1}{2}}n^{-\frac{1}{2}}.$$

First suppose n *odd;* in Art. 262 put for m successively $\frac{1}{n}$, $\frac{2}{n}$, $\frac{3}{n}$, ... up to $\frac{n-1}{2n}$, and multiply; thus

$$\Gamma\left(\frac{1}{n}\right)\Gamma\left(\frac{2}{n}\right)\Gamma\left(\frac{3}{n}\right)\ldots\ldots\Gamma\left(\frac{n-1}{n}\right)=\frac{\pi^{\frac{n-1}{2}}}{\sin\frac{\pi}{n}\sin\frac{2\pi}{n}\ldots\sin\frac{(n-1)\pi}{2n}}$$

$$=(2\pi)^{\frac{n-1}{2}}n^{-\frac{1}{2}}.$$

(See *Plane Trigonometry*, Chap. XXIII.)

Next suppose n *even;* in this case put for m successively $\frac{1}{n}$, $\frac{2}{n}$, ... up to $\frac{n-2}{2n}$, and form the product as before; then multiply the left-hand member by $\Gamma(\frac{1}{2})$ and the right-hand member by the equivalent $\sqrt{\pi}$; then we obtain the same result as before.

267. A still more general formula is

$$\Gamma(x)\,\Gamma\left(x+\frac{1}{n}\right)\Gamma\left(x+\frac{2}{n}\right)\ldots\ldots\Gamma\left(x+\frac{n-1}{n}\right)$$

$$=\Gamma(nx)\,(2\pi)^{\frac{n-1}{2}}n^{\frac{1}{2}-nx},$$

which we shall now prove. Let $\phi(x)$ denote

$$\frac{n^{nx}\Gamma(x)\,\Gamma\left(x+\frac{1}{n}\right)\ldots\ldots\Gamma\left(x+\frac{n-1}{n}\right)}{n\Gamma(nx)};$$

we have then to shew that $\phi(x) = (2\pi)^{\frac{n-1}{2}} n^{-\frac{1}{2}}$.

We have

$$\phi(x+1) = \frac{n^{nx+n}\Gamma(x+1)\,\Gamma\left(x+1+\frac{1}{n}\right)\ldots.\Gamma\left(x+1+\frac{n-1}{n}\right)}{n\Gamma(nx+n)}$$

$$= \frac{n^n x\left(x+\frac{1}{n}\right)\left(x+\frac{2}{n}\right)\ldots\ldots\left(x+\frac{n-1}{n}\right)}{(nx+n-1)(nx+n-2)\ldots\ldots nx}\,\phi(x)$$

$$= \phi(x).$$

Similarly $\phi(x+2) = \phi(x+1) = \phi(x)$; and by proceeding thus we have

$$\phi(x) = \phi(x+m),$$

where m may be as great as we please. Hence $\phi(x)$ is equal to the limit of $\phi(\mu)$ when μ is infinite; thus $\phi(x)$ *must be independent of* x, that is, must have the same value whatever x may be; hence $\phi(x)$ must have the same value as it has when $x=\frac{1}{n}$; thus the theorem follows by the preceding article. This theorem is ascribed to Gauss; a more rigid proof is given in Legendre's *Exercices de Calcul Intégral*, Vol. II. p. 23; see also the *Journal de l'Ecole Polytechnique*, Vol. XVI. p. 212.

268. Many definite integrals may be expressed in terms of the *Gamma-function;* we shall give some examples.

The integral $\int_0^\infty e^{-a^2x^2}\,dx$ becomes by putting y for a^2x^2

$$\int_0 \frac{e^{-y}\,dy}{2a\sqrt{y}}, \text{ that is, } \frac{1}{2a}\Gamma(\tfrac{1}{2}), \text{ or } \frac{\sqrt{\pi}}{2a}.$$

Again, in $\int_0^1 \frac{x^{l-1}(1-x)^{m-1}\,dx}{(x+a)^{l+m}}$ put $\frac{x}{x+a} = \frac{y}{1+a}$; thus we obtain

$$\frac{1}{a^m(1+a)^l}\int_0^1 y^{l-1}(1-y)^{m-1}\,dy, \text{ that is, } \frac{1}{a^m(1+a)^l}\,\frac{\Gamma(l)\,\Gamma(m)}{\Gamma(l+m)}.$$

Again, in $\int_0^1 x^{l-1}(1-x^2)^{m-1}\,dx$ put $x^2 = y$; thus we obtain

$$\frac{1}{2}\int_0^1 y^{\frac{l}{2}-1}(1-y)^{m-1}\,dy, \text{ that is, } \frac{\Gamma\left(\frac{l}{2}\right)\Gamma(m)}{2\Gamma\left(\frac{l}{2}+m\right)}.$$

Thus $$\int_0^{\frac{\pi}{2}} \sin^p\theta\cos^q\theta\,d\theta = \int_0^1 x^p(1-x^2)^{\frac{q-1}{2}}\,dx$$

$$= \int_0^1 x^{p+1-1}(1-x^2)^{\frac{q+1}{2}-1}\,dx = \frac{\Gamma\left(\frac{p+1}{2}\right)\Gamma\left(\frac{q+1}{2}\right)}{2\Gamma\left(\frac{p+q}{2}+1\right)}.$$

Again, in $\int_0^1 \frac{x^{l-1}(1-x)^{m-1}\,dx}{\{ax+b(1-x)\}^{l+m}}$ put $x = \frac{by}{a(1-y)+by}$; thus we obtain

$$\frac{1}{a^l b^m}\int_0^1 y^{l-1}(1-y)^{m-1}\,dy, \text{ that is, } \frac{\Gamma(l)\,\Gamma(m)}{a^l b^m\,\Gamma(l+m)}.$$

269. In $\int_0^a x^{l-1}(a-x)^{m-1}\,dx$ put $x = ay$; thus we obtain

$$a^{l+m-1}\int_0^1 y^{l-1}(1-y)^{m-1}\,dy, \text{ that is, } a^{l+m-1}\frac{\Gamma(l)\,\Gamma(m)}{\Gamma(l+m)}.$$

270. It is required to find the value of the multiple integral

$$\iiint \ldots x^{l-1}\,y^{m-1}\,z^{n-1}\ldots dx\,dy\,dz\ldots$$

the integral being so taken as to give to the variables all *positive* values consistent with the condition that $x+y+z+\ldots$ is not greater than unity.

We will suppose that there are three variables, and consequently that the integral is a triple integral; the method adopted will be seen to be applicable for any number of variables.

We must first integrate for one of the variables, suppose z; the limits then will be 0 and $1-x-y$; thus between these limits

$$\int z^{n-1}\,dz = \frac{(1-x-y)^n}{n} = \frac{\Gamma(n)}{\Gamma(n+1)}(1-x-y)^n.$$

Next integrate with respect to one of the remaining variables, suppose y; the limits will be 0 and $1-x$; and between these limits, by Art. 269,

$$\int y^{m-1}(1-x-y)^n\,dy = \frac{(1-x)^{m+n}\,\Gamma(m)\,\Gamma(n+1)}{\Gamma(m+n+1)}.$$

Lastly integrate with respect to x between the limits 0 and 1; thus between these limits

$$\int x^{l-1}(1-x)^{m+n}\,dx = \frac{\Gamma(l)\,\Gamma(m+n+1)}{\Gamma(l+m+n+1)}.$$

Hence the final result is

$$\frac{\Gamma(n)}{\Gamma(n+1)}\,\frac{\Gamma(m)\,\Gamma(n+1)}{\Gamma(m+n+1)}\,\frac{\Gamma(l)\,\Gamma(m+n+1)}{\Gamma(l+m+n+1)},$$

that is,

$$\frac{\Gamma(l)\,\Gamma(m)\,\Gamma(n)}{\Gamma(l+m+n+1)}.$$

271. It is required to find the value of the multiple integral

$$\iiint\ldots\xi^{l-1}\,\eta^{m-1}\,\zeta^{n-1}\ldots d\xi\,d\eta\,d\zeta\ldots$$

the integral being so taken as to give to the variables all positive values consistent with the condition that

$$\left(\frac{\xi}{\alpha}\right)^p + \left(\frac{\eta}{\beta}\right)^q + \left(\frac{\zeta}{\gamma}\right)^r + \dots$$

is not greater than unity.

Assume $\quad x = \left(\frac{\xi}{\alpha}\right)^p, \quad y = \left(\frac{\eta}{\beta}\right)^q, \quad z = \left(\frac{\zeta}{\gamma}\right)^r, \dots\dots$

Then the integral becomes

$$\frac{\alpha^l \beta^m \gamma^n \dots}{p\,q\,r \dots} \iiint \dots x^{\frac{l}{p}-1}\, y^{\frac{m}{q}-1}\, z^{\frac{n}{r}-1} \dots dx\, dy\, dz \dots$$

with the condition that $x + y + z + \dots$ is not greater than unity. The value of the integral is, therefore, by the preceding article

$$\frac{\alpha^l \beta^m \gamma^n \dots}{p\,q\,r \dots} \frac{\Gamma\left(\frac{l}{p}\right) \Gamma\left(\frac{m}{q}\right) \Gamma\left(\frac{n}{r}\right) \dots\dots}{\Gamma\left(\frac{l}{p} + \frac{m}{q} + \frac{n}{r} + \dots + 1\right)}.$$

272. As a simple case of the preceding article we may suppose $p, q, r, \dots$ to be each unity, and $\alpha, \beta, \gamma, \dots$ each equal to a constant h; thus the condition is that $\xi + \eta + \zeta + \dots$ is not to be greater than h. Therefore the value of the integral

$$\iiint \dots \xi^{l-1} \eta^{m-1} \zeta^{n-1} \dots d\xi\, d\eta\, d\zeta \dots$$

is $$h^{l+m+n+\dots} \frac{\Gamma(l)\, \Gamma(m)\, \Gamma(n) \dots}{\Gamma(l + m + n + \dots + 1)},$$

which we may denote by

$$N h^{l+m+n+\dots}.$$

Similarly if the integral is to be taken so that the sum of the variables shall not exceed $h + \Delta h$, we obtain for the result

$$N (h + \Delta h)^{l+m+n+\dots}.$$

Hence we conclude that the value of the integral extended over all such positive values of the variables as make the sum of the variables lie between h and $h+\Delta h$ is

$$N\{(h+\Delta h)^{l+m+n+\dots} - h^{l+m+n+\dots}\},$$

and when Δh is indefinitely diminished, this becomes

$$N(l+m+n+\dots)\, h^{l+m+n+\dots-1}\Delta h,$$

that is,
$$\frac{\Gamma(l)\,\Gamma(m)\,\Gamma(n)\dots}{\Gamma(l+m+n+\dots)}\, h^{l+m+n+\dots-1}\Delta h.$$

273. It is required to find the value of the multiple integral

$$\iiint\dots x^{l-1}\, y^{m-1}\, z^{n-1}\dots f(x+y+z+\dots)\, dx\, dy\, dz\dots$$

the integral being so taken as to give to the variables all *positive* values consistent with the condition that $x+y+z+\dots$ is not greater than c.

We will suppose for simplicity that there are three variables. By the preceding article if $f(x+y+z)$ were replaced by unity that part of the integral which arises from supposing the sum of the variables to lie between h and $h+\Delta h$ would be ultimately

$$\frac{\Gamma(l)\,\Gamma(m)\,\Gamma(n)}{\Gamma(l+m+n)}\, h^{l+m+n-1}\Delta h.$$

And if the sum of the variables lies between h and $h+\Delta h$, the value of $f(x+y+z)$ can only differ from $f(h)$ by a small quantity of the same order as Δh. Hence, neglecting the square of Δh, that part of the integral which arises from supposing the sum of the variables to lie between h and $h+\Delta h$ is ultimately

$$\frac{\Gamma(l)\,\Gamma(m)\,\Gamma(n)}{\Gamma(l+m+n)} f(h)\, h^{l+m+n-1}\Delta h.$$

Hence the whole integral is

$$\frac{\Gamma(l)\,\Gamma(m)\,\Gamma(n)}{\Gamma(l+m+n)}\int_0^c f(h)\, h^{l+m+n-1}\, dh.$$

274. Similarly the value of

$$\iiint \xi^{l-1} \eta^{m-1} \zeta^{n-1} f\left\{\left(\frac{\xi}{\alpha}\right)^p + \left(\frac{\eta}{\beta}\right)^q + \left(\frac{\zeta}{\gamma}\right)^r\right\} d\xi\, d\eta\, d\zeta$$

for all positive values of the variables, such that

$$\left(\frac{\xi}{\alpha}\right)^p + \left(\frac{\eta}{\beta}\right)^q + \left(\frac{\zeta}{\gamma}\right)^r$$

is not greater than c, is

$$\frac{\alpha^l \beta^m \gamma^n}{p\,q\,r} \frac{\Gamma\left(\frac{l}{p}\right)\Gamma\left(\frac{m}{q}\right)\Gamma\left(\frac{n}{r}\right)}{\Gamma\left(\frac{l}{p}+\frac{m}{q}+\frac{n}{r}\right)} \int_0^c f(h)\, h^{\frac{l}{p}+\frac{m}{q}+\frac{n}{r}-1}\, dh.$$

The result of this and the preceding article may be extended to the case of any number of variables.

275. It is required to find the value of the multiple integral

$$\iiint \ldots f(a_1x_1 + a_2x_2 + \ldots\ldots + a_nx_n)\, dx_1\, dx_2 \ldots dx_n,$$

the integral being so taken as to give to the variables *all* values consistent with the condition that $x_1^2 + x_2^2 \ldots + x_n^2$ is not greater than unity.

By successive application of a transformation for a double integral given in Art. 242, the multiple integral may be reduced to

$$\iiint \ldots f(kx_1)\, dx_1\, dx_2 \ldots dx_n,$$

where $$k = \sqrt{(a_1^2 + a_2^2 + \ldots + a_n^2)};$$

and these transformations do not affect the condition that the sum of the squares of the variables is not to be greater than unity.

We have first then to find the value of the multiple integral $\iiint \ldots dx_2\, dx_3 \ldots dx_n$, the variables being supposed to have all values consistent with the condition that $x_2^2 + x_3^2 + \ldots + x_n^2$ is not greater than $1 - x_1^2$. First suppose that the variables are to have only *positive* values; then we obtain the value of the integral by supposing in Art. 271, that each of the quantities $l, m, \ldots$ is unity, that each of the quantities $p, q, \ldots$ is equal to 2, and that each of the quantities $\alpha, \beta, \ldots$ is equal to $\sqrt{(1 - x_1^2)}$. Thus the result is

$$\frac{\{\Gamma(\frac{1}{2})\}^{n-1}}{2^{n-1}\Gamma\left(\frac{n-1}{2}+1\right)}(1-x_1^2)^{\frac{n-1}{2}}.$$

But if the variables may have negative as well as positive values, this result must be multiplied by 2^{n-1}. Thus we get

$$\frac{\pi^{\frac{n-1}{2}}(1-x_1^2)^{\frac{n-1}{2}}}{\Gamma\left(\frac{n-1}{2}+1\right)}.$$

Hence, finally, since the limits of x_1 will be -1 and 1, the multiple integral is equal to

$$\frac{\pi^{\frac{n-1}{2}}}{\Gamma\left(\frac{n-1}{2}+1\right)}\int_{-1}^{1} f(kx_1)(1-x_1^2)^{\frac{n-1}{2}}\, dx_1.$$

This agrees with the result given by Professor Boole in the *Cambridge Mathematical Journal*, Vol. III. p. 280, as it may be found by integrating his equation (15) by parts.

276. It is required to find the value of the multiple integral

$$\iiint \ldots \frac{f(a_1x_1 + a_2x_2 + \ldots + a_nx_n)}{\sqrt{(1 - x_1^2 - x_2^2 \ldots - x_n^2)}}\, dx_1\, dx_2 \ldots dx_n,$$

the integral being so taken as to give to the variables *all* values consistent with the condition that $x_1^2 + x_2^2 + \ldots + x_n^2$ is not greater than unity.

As in the preceding article the integral may be transformed into

$$\iiint \ldots \frac{f(kx_1)}{\sqrt{(1 - x_1^2 - x_2^2 \ldots - x_n^2)}} dx_1 \, dx_2 \ldots dx_n.$$

First integrate with respect to the variables $x_2, x_3, \ldots x_n$, the limits being given by the condition that $x_2^2 + x_3^2 \ldots + x_n^2$ is not greater than $1 - x_1^2$. Now if the variables were restricted to positive values, the integral

$$\iiint \ldots \frac{dx_2 \, dx_3 \ldots dx_n}{\sqrt{(1 - x_1^2 - x_2^2 \ldots - x_n^2)}}$$

by Art. 274 would be equal to

$$\frac{1}{2^{n-1}} \frac{\{\Gamma(\frac{1}{2})\}^{n-1}}{\Gamma\left(\frac{n-1}{2}\right)} \int_0^{1-x_1^2} (1 - x_1^2 - h)^{-\frac{1}{2}} h^{\frac{n-1}{2}-1} \, dh,$$

that is, to

$$\frac{1}{2^{n-1}} \frac{\{\Gamma(\frac{1}{2})\}^{n-1}}{\Gamma\left(\frac{n-1}{2}\right)} (1 - x_1^2)^{\frac{n}{2}-1} \frac{\Gamma(\frac{1}{2}) \, \Gamma\left(\frac{n-1}{2}\right)}{\Gamma\left(\frac{n}{2}\right)}, \qquad \text{(Art. 269)},$$

that is, to

$$\frac{1}{2^{n-1}} \frac{\{\Gamma(\frac{1}{2})\}^{n}}{\Gamma\left(\frac{n}{2}\right)} (1 - x_1^2)^{\frac{n}{2}-1}.$$

But if the variables may have negative as well as positive values, this result must be multiplied by 2^{n-1}. Thus we get

$$\frac{\pi^{\frac{n}{2}}}{\Gamma\left(\frac{n}{2}\right)} (1 - x_1^2)^{\frac{n}{2}-1}.$$

Hence finally, since the limits of x_1 are -1 and 1, the multiple integral is equal to

$$\frac{\pi^{\frac{n}{2}}}{\Gamma\left(\frac{n}{2}\right)}\int_{-1}^{1} f(kx_1)\,(1-x_1^2)^{\frac{n}{2}-1}\,dx_1.$$

277. Many methods have been used for exhibiting in simple terms an approximate value of $\Gamma(n+1)$ when n is very large: we give one of them.

The product $e^{-x}x^n$ vanishes when $x=0$ and when $x=\infty$; and it may be shewn that it has only one maximum value, namely when $x=n$. We may therefore assume

$$e^{-x}x^n = e^{-n}n^n e^{-t^2} \text{.................... (1)},$$

where t is a variable which must lie between the limits $-\infty$ and $+\infty$.

Thus $$\int_0^\infty e^{-x}x^n\,dx = e^{-n}n^n\int_{-\infty}^{\infty} e^{-t^2}\frac{dx}{dt}\,dt \text{ .... (2)}.$$

Take the logarithms of both members of (1); thus

$$x - n\log x = n - n\log n + t^2 \text{.............. (3)};$$

put $x = n+u$; thus

$$u - n\log(n+u) = t^2 - n\log n \text{ .... (4)}.$$

But by Taylor's Theorem

$$\log(n+u) = \log n + \frac{u}{n} - \frac{u^2}{2(n+\theta u)^2},$$

where θ is a proper fraction; thus (4) becomes

$$\frac{nu^2}{2(n+\theta u)^2} = t^2;$$

therefore $$\frac{\sqrt{(n)}\,u}{\sqrt{(2)}\,(n+\theta u)} = t \text{ ....(5)};$$

therefore $$u = \frac{\sqrt{(2)}\, nt}{\sqrt{(n)} - \theta t \sqrt{2}} \qquad\qquad (6).$$

But from (3) $$\frac{dx}{dt} = \frac{2xt}{x-n} = 2t + \frac{2nt}{u}$$

$$= \sqrt{(2n)} + 2(1-\theta)\,t, \qquad \text{by (6).}$$

Hence (2) becomes

$$\int_0^\infty e^{-x} x^n\, dx = e^{-n} n^n \int_{-\infty}^\infty e^{-t^2} \{\sqrt{(2n)} + 2(1-\theta)\,t\}\, dt\,;$$

and $\int_{-\infty}^\infty e^{-t^2} dt = \sqrt{(\pi)}$; thus

$$\int_0^\infty e^{-x} x^n dx = e^{-n} n^n \sqrt{(2n\pi)}\, \{1 + \frac{2}{\sqrt{(2n\pi)}} \int_{-\infty}^\infty e^{-t^2} (1-\theta)\, t\, dt\} \quad (7).$$

But since $1-\theta$ is positive and less than unity, the numerical value of $\int_{-\infty}^\infty e^{-t^2} (1-\theta)\, t\, dt$ is less than $\int_0^\infty e^{-t^2} t\, dt$, that is, less than $\frac{1}{2}$. Hence we conclude from (7) that as n is increased indefinitely, the ratio of $\Gamma(n+1)$ to $e^{-n} n^n \sqrt{(2n\pi)}$ approaches unity as its limit.

We may observe that in the original equation (1) we have t^2 and not t itself; hence the sign of t is in our power, and we accordingly take it so that equation (5) may hold, supposing $\sqrt{n}$ and $\sqrt{2}$ both positive.

(See Liouville's *Journal de Mathématiques*, Vol. X. p. 464, and Vol. XVII. p. 448.)

Definite Integrals obtained by differentiating or integrating with respect to constants.

278. We shall now give some examples in which definite integrals are obtained by means of differentiation with respect to a constant. (See Art. 213.)

To find the value of $\int_0^\infty e^{-a^2x^2} \cos 2rx \, dx$. Call the definite integral u; then

$$\frac{du}{dr} = -2\int_0^\infty xe^{-a^2x^2} \sin 2rx \, dx.$$

Integrate the right-hand term by parts; thus we find

$$\frac{du}{dr} = -\frac{2ru}{a^2};$$

therefore
$$\frac{d \log u}{dr} = -\frac{2r}{a^2};$$

therefore
$$\log u = -\frac{r^2}{a^2} + \text{constant},$$

therefore
$$u = Ae^{-\frac{r^2}{a^2}},$$

where A is a quantity which is constant with respect to r, that is, it does not contain r. To determine A we may suppose $r = 0$; thus u becomes $\int_0^\infty e^{-a^2x^2} \, dx$, that is, $\frac{\sqrt{\pi}}{2a}$, (Art. 268).

Hence $A = \frac{\sqrt{\pi}}{2a}$, and

$$\int_0^\infty e^{-a^2x^2} \cos 2rx \, dx = \frac{\sqrt{\pi}}{2a} e^{-\frac{r^2}{a^2}}.$$

279. We have stated in Art. 214, that when one of the limits of integration is *infinite* the process of differentiation with respect to a constant *may* be unsafe; in the present case however it is easy to justify it; we have to shew that $\int_0^\infty e^{-a^2x^2} \rho \, dx$ vanishes where ρ is ultimately indefinitely small; it is obvious that this quantity is numerically less than $\rho_1 \int_0^\infty e^{-a^2x^2} \, dx$ where ρ_1 is the greatest value of ρ, that is,

less than $\frac{\sqrt{\pi}}{2a}\rho_1$; but this vanishes since ρ_1 does. Similar considerations apply to the succeeding cases.

280. To find the value of $\int_0^\infty e^{-kx}\frac{\sin rx\,dx}{x}$. Denote it by u, then

$$\frac{du}{dr}=\int_0^\infty e^{-kx}\cos rx\,dx.$$

But $$\int e^{-kx}\cos rx\,dx=e^{-kx}\frac{r\sin rx-k\cos rx}{k^2+r^2};$$

therefore $$\int_0^\infty e^{-kx}\cos rx\,dx=\frac{k}{k^2+r^2};$$

thus $$\frac{du}{dr}=\frac{k}{k^2+r^2};$$

therefore $$u=\tan^{-1}\frac{r}{k}.$$

No constant is required because u vanishes with r. This result holds for any positive value of k; if we suppose k to diminish without limit, we obtain

$$\int_0^\infty \frac{\sin rx}{x}\,dx=\frac{\pi}{2}$$

if r be positive; if r be negative the result should be $-\frac{\pi}{2}$.

281. To find the value of $\int_0^\infty e^{-\left(x^2+\frac{a^2}{x^2}\right)}dx$. Denote it by u, then

$$\frac{du}{da}=-2a\int_0^\infty e^{-\left(x^2+\frac{a^2}{x^2}\right)}\frac{dx}{x^2};$$

assume $x=\frac{a}{z}$, then the limits of z are ∞ and 0; and we obtain

$$\frac{du}{da}=-2u;$$

therefore $$\frac{d \log u}{da} = -2 ;$$

therefore $$\log u = -2a + \text{constant};$$

therefore $$u = Ae^{-2a}.$$

To determine A we may suppose $a = 0$; then $u = \frac{\sqrt{\pi}}{2}$; therefore $A = \frac{\sqrt{\pi}}{2}$; thus

$$\int_0^\infty e^{-\left(x^2 + \frac{a^2}{x^2}\right)} dx = \frac{\sqrt{\pi}}{2} e^{-2a}.$$

282. We may also apply the principle of *integration* with respect to a constant in order to determine some definite integrals; the principle may be established thus.

Let $$u = \int_a^b \phi(x, c)\, dx,$$

then $$\int_\alpha^\beta u dc = \int_\alpha^\beta \int_a^b \phi(x, c)\, dc\, dx$$

$$= \int_a^b \int_\alpha^\beta \phi(x, c)\, dx\, dc;$$

since when the limits are constant, the order of integration is indifferent (Art. 62). We shall now give some examples of this method.

283. We know that $\int_0^\infty e^{-kx}\, dx = \frac{1}{k}$.

Integrate both sides with respect to k between the limits a and b; thus

$$\int_0^\infty \frac{e^{-ax} - e^{-bx}}{x}\, dx = \log \frac{b}{a}.$$

It should be noticed that $\int_0^\infty \frac{e^{-ax}\, dx}{x}$ and $\int_0^\infty \frac{e^{-bx}\, dx}{x}$ are both infinite; for $\int_0^c \frac{e^{-ax}\, dx}{x}$ is greater than $e^{-ca} \int_0^c \frac{dx}{x}$, and $\int_0^c \frac{dx}{x}$

is infinite. But this is not inconsistent with the assertion that $\int_0^\infty \frac{e^{-ax} - e^{-bx}}{x} dx$ is finite, and without finding the value of this integral it is easy to shew that it must be finite. For it is equal to the sum of $\int_0^c \frac{\phi(x)\,dx}{x}$ and $\int_c^\infty \frac{\phi(x)\,dx}{x}$ where $\phi(x) = e^{-ax} - e^{-bx}$; the second of these integrals is finite, for it is less than $\frac{1}{c}\int_c^\infty \phi(x)\,dx$, that is, less than $\frac{1}{c}\left(\frac{e^{-ac}}{a} - \frac{e^{-bc}}{b}\right)$.

We have then only to examine $\int_0^c \frac{\phi(x)}{x} dx$.

Now by Maclaurin's Theorem

$$\phi(x) = (b-a)\,x + \frac{x^2}{2}\phi''(x\theta),$$

where θ is some fraction; thus $\frac{\phi(x)}{x}$ is less than $b - a + \frac{Ax}{2}$, where A is the greatest value which $\phi''(x)$ can assume for values of x less than c. Hence

$$\int_0^c \frac{\phi(x)}{x} dx \text{ is less than } (b-a)\,c + \frac{Ac^2}{4},$$

and is therefore finite.

284. We know that

$$\int_0^\infty e^{-kx} \cos rx\,dx = \frac{k}{k^2 + r^2}.$$

Integrate both sides with respect to k between the limits a and b; thus

$$\int_0^\infty \frac{e^{-ax} - e^{-bx}}{x} \cos rx\,dx = \tfrac{1}{2}\log\frac{b^2 + r^2}{a^2 + r^2}.$$

285. Let $\int_0^\infty \frac{\sin rx}{x} dx$ be denoted by A, and $\int_0^\infty \frac{\cos rx}{1+x^2} dx$ by B; we shall now determine the values of A and B; the former has already been determined by another method in Art. 280.

In the integral A put y for rx; thus

$$A = \int_0^\infty \frac{\sin y \, dy}{y};$$

this shews that A is independent of r.

We have $$\frac{dB}{dr} = -\int_0^\infty \frac{x \sin rx \, dx}{1+x^2},$$

and $$\int_0^r B dr = \int_0^\infty \frac{\sin rx \, dx}{x(1+x^2)};$$

thus $$\int_0^r B dr - \frac{dB}{dr} = \int_0^\infty \frac{1+x^2}{x} \frac{\sin rx}{1+x^2} dx = A;$$

hence $$\int_0^r B dr - \frac{dB}{dr} - A = 0 \quad\ldots\ldots\ldots\ldots\ldots (1).$$

Multiply by e^{-r} and integrate; we obtain since A is constant with respect to r

$$e^{-r}\left\{\int_0^r B dr + B - A\right\} = \text{constant}.$$

Now whatever be the value of r, it is obvious that the integrals represented by A, B, and $\int_0^r B dr$ are *finite*; hence the constant in the last equation must be zero, for the left-hand member vanishes when r is infinite.

Thus $$\int_0^r B dr + B - A = 0 \quad\ldots\ldots\ldots\ldots\ldots\ldots\ldots (2).$$

From (1) and (2) $$\frac{dB}{dr} = -B;$$

therefore $$B = Ce^{-r},$$

where C is some constant. And from (2)

$$A = Ce^{-r} - C(e^{-r} - 1) = C;$$

therefore $$B = Ae^{-r} \quad\ldots\ldots\ldots\ldots\ldots\ldots\ldots\ldots (3).$$

Now when r is indefinitely diminished, B becomes $\int_0^\infty \frac{dx}{1+x^2}$, that is, $\frac{\pi}{2}$; hence from (3)

$$A = \frac{\pi}{2} \text{ and } B = \frac{\pi}{2} e^{-r}.$$

We have supposed r positive; it is obvious that if r be negative, B has the same value as if r were positive, and A has its sign changed; that is, if r be negative $B = \frac{\pi}{2} e^r$ and $A = -\frac{\pi}{2}$. (*Transactions of the Royal Irish Academy*, Vol. XIX. p. 277.)

From $\int_0^\infty \frac{\cos rx\, dx}{1+x^2} = \frac{\pi}{2} e^{-r}$, we obtain by differentiation with respect to r,

$$\int_0^\infty \frac{x \sin rx\, dx}{1+x^2} = \frac{\pi}{2} e^{-r}.$$

And from the same integral by integrating with respect to r between the limits 0 and c, we have

$$\int_0^\infty \frac{\sin cx\, dx}{x(1+x^2)} = \frac{\pi}{2}(1-e^{-c}).$$

286. The preceding article contains a rigorous investigation of the values of the integrals A and B; another method has been sometimes given for finding the value of B which is more simple but far less satisfactory. We will however now give this method, as it will lead us to notice a point of importance.

Let $$B = \int_0^\infty \frac{\cos rx}{1+x^2}\, dx,$$

then $$\frac{dB}{dr} = -\int_0^\infty \frac{x \sin rx}{1+x^2}\, dx,$$

and $$\frac{d^2B}{dr^2} = -\int_0^\infty \frac{x^2 \cos rx}{1+x^2}\,dx$$

$$= -\int_0^\infty \cos rx\,dx + \int_0^\infty \frac{\cos rx}{1+x^2}\,dx$$

$$= -\int_0^\infty \cos rx\,dx + B.$$

Now we will *assume* on grounds presently to be examined, that $\int_0^\infty \cos rx\,dx = 0$; thus

$$\frac{d^2B}{dr^2} = B,$$

and we have to find B from this equation. Multiply both sides by $2\frac{dB}{dr}$ and integrate with respect to r; thus

$$\left(\frac{dB}{dr}\right)^2 = h + B^2;$$

where h is a constant, that is, h is independent of r. Thus

$$\frac{dB}{dr} = \sqrt{(h+B^2)},$$

therefore $$\frac{dr}{dB} = \frac{1}{\sqrt{(h+B^2)}};$$

by integrating we have

$$r + k = \int \frac{dB}{\sqrt{(h+B^2)}} = \log\{B + \sqrt{(h+B^2)}\},$$

where k is another constant.

Thus $$e^{r+k} = B + \sqrt{(h+B^2)}.$$

By transposing, squaring, and reducing we finally obtain

$$B = C_1 e^r + C_2 e^{-r},$$

where C_1 and C_2 are constants. We must now determine the values of these constants. Since B cannot increase indefinitely with r we must have $C_1=0$; and then since $B=\frac{\pi}{2}$ when $r=0$ we have $C_2=\frac{\pi}{2}$. Thus

$$B=\frac{\pi}{2}e^{-r}.$$

We now proceed to consider the assumption involved in the preceding method.

Since $\int e^{-ax}\sin rx\,dx=-e^{-ax}\frac{a\sin rx+r\cos rx}{a^2+r^2}$,

and $\int e^{-ax}\cos rx\,dx=e^{-ax}\frac{r\sin rx-a\cos rx}{a^2+r^2}$,

we have $\int_0^\infty e^{-ax}\sin rx\,dx=\frac{r}{a^2+r^2}$,

and $\int_0^\infty e^{-ax}\cos rx\,dx=\frac{a}{a^2+r^2}$,

if a be a positive quantity.

If it were allowable to suppose $a=0$ we should obtain

$$\int_0^\infty \sin rx\,dx=\frac{1}{r},\ \text{and}\int_0^\infty \cos rx\,dx=0.$$

Since $\int \sin rx\,dx=-\frac{\cos rx}{r}$, and $\int \cos rx\,dx=\frac{\sin rx}{r}$, we are thus apparently led to the conclusion that *the sine and cosine of an infinite angle are both zero.* The same conclusion seems to be suggested in other cases, so that it has been stated, that "the indeterminate symbols $\sin\infty$ and $\cos\infty$ are found in numberless cases to represent each of them, 0, the mean value of both $\sin x$ and $\cos x$."

On this point however diversity of opinion exists among mathematicians, and the discussion of it would be unsuitable to an elementary work; the student may hereafter consult three memoirs in the eighth volume of the *Cambridge Philosophical Transactions*, numbered XV, XIX, and XXXII.

Definite Integrals obtained by Expansion.

287. If we expand $\log \{1-ae^{x\sqrt{(-1)}}\}$ and $\log \{1-ae^{-x\sqrt{(-1)}}\}$ and add, we obtain

$$\log (1-2a \cos x+a^2)$$
$$=-2\left(a \cos x+\frac{a^2}{2} \cos 2x+\frac{a^3}{3} \cos 3x+\ldots\ldots\right),$$

the series being convergent if a is less than unity. Integrate both sides with respect to x between the limits 0 and π; thus

$$\int_0^\pi \log (1-2a \cos x+a^2)\,dx=0, \ a \text{ being less than } 1.$$

If a is greater than 1, since

$$\log (1-2a \cos x+a^2)=\log a^2+\log \left(1-\frac{2}{a} \cos x+\frac{1}{a^2}\right),$$

we have

$$\int_0^\pi \log (1-2a \cos x+a^2)\,dx=\pi \log a^2=2\pi \log a.$$

If $a=1$ it may be shewn by Art. 51 that the definite integral is zero.

We may put the result in the following form;

$$\int_0^\pi \log (a^2-2ac \cos x+c^2)\,dx=\pi \log k^2,$$

where k^2 is the greater of the two quantities a^2 and c^2, and is equal to either of them if they are equal.

By differentiating this result with respect to a we arrive at the result which constitutes the last example of Art. 50.

288. By integration by parts we have

$$\int \log(1-2a\cos x+a^2)\,dx$$

$$=x\log(1-2a\cos x+a^2)-2a\int\frac{x\sin x\,dx}{1-2a\cos x+a^2}.$$

Hence, if a be less than 1,

$$\int_0^\pi \frac{x\sin x\,dx}{1-2a\cos x+a^2}=\frac{\pi}{2a}\log(1+a)^2, \text{ that is, } \frac{\pi}{a}\log(1+a);$$

if a be greater than 1, the result is

$$\frac{\pi}{a}\log(1+a)-\frac{\pi}{a}\log a, \text{ that is, } \frac{\pi}{a}\log\left(1+\frac{1}{a}\right).$$

289. In like manner we have, if r be an integer,

$$\int_0^\pi \cos rx\log(1-2a\cos x+a^2)\,dx=-\frac{\pi}{r}a^r, \text{ or } -\frac{\pi}{r}a^{-r},$$

according as a is less or greater than unity.

290. Integrate by parts the integral in the preceding article; thus we find

$$\int_0^\pi \frac{\sin x\sin rx\,dx}{1-2a\cos x+a^2}=\frac{\pi}{2}a^{r-1} \text{ or } \frac{\pi}{2}a^{-(r+1)},$$

according as a is less or greater than unity.

291. Similarly from the known expansion

$$\frac{1-a^2}{1-2a\cos x+a^2}$$

$$=1+2a\cos x+2a^2\cos 2x+2a^3\cos 3x+\ldots\ldots,$$

where a is less than 1, we may deduce some definite integrals; thus if r is an integer

$$\int_0^\pi \frac{\cos rx\,dx}{1-2a\cos x+a^2}=\frac{\pi a^r}{1-a^2},$$

for every term that we have to integrate vanishes with the assigned limits, except $2a^r\int_0^\pi \cos^2 rx\,dx$.

292. To find the value of $\int_0^\infty \frac{1}{1+x^2}\,\frac{dx}{1-2a\cos cx+a^2}$.

The term $\frac{1}{1-2a\cos cx+a^2}$ may be expanded as in Art. 291; then each term may be integrated by Art. 286, and the results summed. Thus we shall obtain

$$\frac{\pi}{2}\cdot\frac{1}{1-a^2}\,\frac{1+ae^{-c}}{1-ae^{-c}}.$$

293. Similarly,

$$\int_0^\infty \log(1-2a\cos cx+a^2)\,\frac{dx}{1+x^2}=\pi\log(1-ae^{-c}).$$

294. It is also known from Trigonometry that

$$\frac{\sin cx}{1-2a\cos cx+a^2}=\sin cx+a\sin 2cx+a^2\sin 3cx+\dots,$$

a being less than 1. Hence by Art. 286, we obtain

$$\int_0^\infty \frac{x\sin cx\,dx}{(1+x^2)(1-2a\cos cx+a^2)}=\frac{\pi}{2(e^c-a)}.$$

This also follows from Art. 293, by differentiating with respect to c.

295. To find $\int_0^1 \frac{\log x}{1-x}\,dx$.

By expanding $(1-x)^{-1}$, we find for the integral a series of which the type is

$$\int_0^1 x^n\log x\,dx.$$

By integration by parts this is seen to be equal to $-\frac{1}{(1+n)^2}$. Hence the result is

$$-\left\{1+\frac{1}{2^2}+\frac{1}{3^2}+\frac{1}{4^2}+\ldots\right\},$$

that is, by a known formula, $-\frac{\pi^2}{6}$.

296. To find $\int_0^\pi \frac{x \sin x \, dx}{1+(\cos x)^2}$.

Expand the factor $\{1+(\cos x)^2\}^{-1}$, and we find for the integral a series of which the type is

$$(-1)^n \int_0^\pi x \sin x (\cos x)^{2n} \, dx.$$

By integration by parts this may be shewn to be equal to $\frac{(-1)^n \pi}{2n+1}$.

Hence the result is

$$\pi\left\{1-\frac{1}{3}+\frac{1}{5}-\frac{1}{7}+\ldots\right\},$$

that is, by a known formula, $\frac{\pi^2}{4}$.

297. Let v denote $e^{x\sqrt{(-1)}}$, that is, $\cos x+\sqrt{(-1)} \sin x$; then if f denote any function, we have by Taylor's Theorem,

$$f(a+v)+f(a+v^{-1})$$

$$=2\left\{f(a)+f'(a)\cos x+\frac{f''(a)}{1.2}\cos 2x+\ldots\ldots\right\}.$$

And

$$\frac{1-c^2}{1-2c\cos x+c^2}=1+2c\cos x+2c^2\cos 2x+2c^3\cos 3x+\ldots\ldots$$

Therefore

$$\int_0^\pi \frac{f(a+v)+f(a+v^{-1})}{1-2c\cos x+c^2}\,dx=\frac{2\pi}{1-c^2}\left\{f(a)+cf'(a)+\frac{c^2}{1.2}f''(a)+\ldots\right\}$$

$$=\frac{2\pi}{1-c^2}f(a+c).$$

In this result it must be remembered that c is to be less than unity, and the functions $f(a+v)$ and $f(a+v^{-1})$ must be such that Taylor's Theorem holds for their expansions.

In a similar way it may be shewn that

$$\int_0^\pi \frac{f(a+v)-f(a+v^{-1})}{1-2c\cos x+c^2}\sin x\,dx=\frac{\pi\sqrt{(-1)}}{c}\{f(a+c)-f(a)\},$$

and $$\int_0^\pi \frac{1-c\cos x}{1-2c\cos x+c^2}\{f(a+v)+f(a+v^{-1})\}\,dx$$

$$=\pi\{f(a+c)+f(a)\}.$$

Substitution of imaginary values for Constants.

298. Definite integrals are sometimes deduced from known integrals by substituting impossible values for some of the constants which occur. This process cannot be considered demonstrative, but will serve at least to suggest the forms which can be examined, and perhaps verified by other methods (see De Morgan's *Differential and Integral Calculus*, p. 630). We will give some examples of it.

We have $\int_0^\infty e^{-px}x^{n-1}\,dx=p^{-n}\Gamma(n).$

For p put $a+b\sqrt{(-1)}$, and suppose $r=\sqrt{(a^2+b^2)}$ and $\tan\theta=\frac{b}{a}$, so that $p=r\{\cos\theta+\sqrt{(-1)}\sin\theta\}$; thus

$$\int_0^\infty e^{-\{a+b\sqrt{(-1)}\}x}x^{n-1}\,dx=r^{-n}\{\cos n\theta-\sqrt{(-1)}\sin n\theta\}\Gamma(n).$$

Thus by separating the possible and impossible parts we have

$$\int_0^\infty e^{-ax} x^{n-1} \cos bx \, dx = \frac{\Gamma(n) \cos\left(n \tan^{-1} \frac{b}{a}\right)}{(a^2 + b^2)^{\frac{n}{2}}}.$$

$$\int_0^\infty e^{-ax} x^{n-1} \sin bx \, dx = \frac{\Gamma(n) \sin\left(n \tan^{-1} \frac{b}{a}\right)}{(a^2 + b^2)^{\frac{n}{2}}}.$$

For modes of verification see De Morgan, p. 630.

299. In the formula

$$\int_0^\infty e^{-a^2x^2} \, dx = \frac{\sqrt{\pi}}{2a}$$

change a into $\dfrac{1 + \sqrt{(-1)}}{\sqrt{2}} c$; thus

$$\int_0^\infty e^{-c^2x^2\sqrt{(-1)}} \, dx = \frac{1 - \sqrt{(-1)}}{2c} \frac{\sqrt{\pi}}{\sqrt{2}};$$

therefore

$$\int_0^\infty \left\{\cos c^2x^2 - \sqrt{(-1)} \sin c^2x^2\right\} dx = \frac{1 - \sqrt{(-1)}}{2c} \frac{\sqrt{\pi}}{\sqrt{2}};$$

therefore $$\int_0^\infty \cos c^2x^2 \, dx = \frac{\sqrt{\pi}}{2c\sqrt{2}},$$

and $$\int_0^\infty \sin c^2x^2 \, dx = \frac{\sqrt{\pi}}{2c\sqrt{2}}.$$

If we write y for c^2x^2, these become

$$\int_0^\infty \frac{\sin y \, dy}{\sqrt{y}} = \int_0^\infty \frac{\cos y \, dy}{\sqrt{y}} = \sqrt{\frac{\pi}{2}}.$$

300. In the integral $\int_0^\infty e^{-\left(x^2+\frac{a^2}{x^2}\right)k}\,dx$, suppose $y = x\sqrt{k}$; thus the integral becomes $\frac{1}{\sqrt{k}}\int_0^\infty e^{-\left(y^2+\frac{k^2a^2}{y^2}\right)}\,dy$, which is known by Art. 281. Thus

$$\int_0^\infty e^{-\left(x^2+\frac{a^2}{x^2}\right)k}\,dx = \frac{1}{\sqrt{k}}\frac{\sqrt{\pi}}{2}e^{-2ak}.$$

Now put $\cos\theta + \sqrt{(-1)}\sin\theta$ for k; thus the right-hand member becomes

$$\frac{1}{\cos\frac{\theta}{2} + \sqrt{(-1)}\sin\frac{\theta}{2}}\cdot\frac{\sqrt{\pi}}{2}e^{-2a\{\cos\theta+\sqrt{(-1)}\sin\theta\}},$$

that is,

$$\frac{\sqrt{\pi}}{2}\left\{\cos\left(2a\sin\theta+\frac{\theta}{2}\right) - \sqrt{(-1)}\sin\left(2a\sin\theta+\frac{\theta}{2}\right)\right\}e^{-2a\cos\theta}.$$

Thus $\int_0^\infty e^{-\left(x^2+\frac{a^2}{x^2}\right)\cos\theta}\cos\left\{\left(x^2+\frac{a^2}{x^2}\right)\sin\theta\right\}dx$

$$= \frac{\sqrt{\pi}}{2}e^{-2a\cos\theta}\cos\left(2a\sin\theta+\frac{\theta}{2}\right),$$

and $\int_0^\infty e^{-\left(x^2+\frac{a^2}{x^2}\right)\cos\theta}\sin\left\{\left(x^2+\frac{a^2}{x^2}\right)\sin\theta\right\}dx$

$$= \frac{\sqrt{\pi}}{2}e^{-2a\cos\theta}\sin\left(2a\sin\theta+\frac{\theta}{2}\right).$$

EXAMPLES.

1. Evaluate $\int_0^\infty \frac{(x^2+a^2)\,dx}{x^4+b^2x^2+b^4}$. *Result.* $\frac{(a^2+b^2)\,\pi}{2b^3\sqrt{3}}$.

2. Evaluate $\int_0^{\frac{1}{2}\pi}\cos(a\tan x)\,dx$. *Result.* $\frac{\pi}{2}e^{-a}$.

3. Evaluate $\int_0^1 x^{2n-1} e^{x^n}\, dx.$ *Result.* $\frac{1}{n}$.

4. $$\int_0^{\frac{\pi}{2}} \frac{dx}{(a^2\cos^2 x + b^2 \sin^2 x)^2} = \frac{\pi}{4}\left(\frac{1}{ab^3} + \frac{1}{a^3 b}\right).$$

5. Prove $\int_0^{\frac{\pi}{4}} \sqrt{(\tan\phi)}\, d\phi = \frac{1}{\sqrt{2}}\left[\frac{\pi}{2} + \log\{\sqrt{(2)} - 1\}\right].$

6. Prove $\int_0^{\frac{\pi}{4}} \sqrt{(\cot\phi)}\, d\phi = \frac{1}{\sqrt{2}}\left[\frac{\pi}{2} + \log\{\sqrt{(2)} + 1\}\right].$

7. Find the limiting value of $xe^{-x^2}\int_0^x e^{x^2}\, dx$ when $x = \infty$.

Result. $\frac{1}{2}$.

8. Shew that $\int_0^\infty \frac{\cos ax - \cos bx}{x}\, dx = \log\frac{b}{a}.$

9. If $F\left(x, \frac{1}{x}\right)$ be any symmetrical function of x and $\frac{1}{x}$, then

$$\int_0^\infty \frac{dx}{xF\left(x, \frac{1}{x}\right)} = 2\int_0^1 \frac{dx}{xF\left(x, \frac{1}{x}\right)}.$$

10. If $F(x)$ be an algebraical polynomial of less than n dimensions

$$\int_b^a \frac{F(x)\, dx}{(x-c)^n} = \frac{1}{\lfloor n-1} \frac{d^{n-1}}{dc^{n-1}}\left\{F(c)\log\frac{a-c}{b-c}\right\}.$$

11. Prove that $\int_0^{2\pi} e^{\cos\theta}\cos(\sin\theta)\, d\theta = 2\pi.$

12. Prove that $\int_0^{\frac{\pi}{2}} \frac{\sqrt{(1-c)}\, d\theta}{1 - c\cos^n\theta} = \frac{\pi}{\sqrt{(2n)}}$ when c is indefinitely nearly equal to unity, n being a positive quantity.

13. Evaluate $\int_0^\pi (a \cos\theta + b \sin\theta) \log (a \cos^2\theta + b \sin^2\theta)\, d\theta$.

Result. $2b \left\{\log a - 2 + \frac{2\sqrt{b}}{\sqrt{(a-b)}} \cos^{-1} \frac{\sqrt{b}}{\sqrt{a}}\right\}$, supposing a greater than b.

14. Shew that

$$\int_0^\infty \log \frac{1 + 2n \cos ax + n^2}{1 + 2n \cos bx + n^2} \cdot \frac{dx}{x} = \log (1 + n) \log \frac{b^2}{a^2},$$

or $\log\left(1 + \frac{1}{n}\right) \log \frac{b^2}{a^2}$, according as n is less or greater than unity.

15. Find the value of

$$\int_0^\infty \{e^{-ax - \alpha x\sqrt{(-1)}} - e^{-bx - \beta x\sqrt{(-1)}}\} \frac{dx}{x},$$

where a and b are positive, but α and β positive or negative; and shew that it is wholly real when $\frac{\alpha}{a} = \frac{\beta}{b}$.

16. Prove that $\int_0^1 \cot^{-1} (1 - x + x^2)\, dx = \frac{\pi}{2} - \log 2$.

17. Prove that $\int_0^\infty \frac{dx}{1+x^2} \log \left(x + \frac{1}{x}\right) = \pi \log 2$.

18. From the value of $\int_0^\infty \frac{\sin x}{x} dx$ deduce that of

$$\int_0^\infty \left(\frac{\sin x}{x}\right)^2 dx.$$

Result. The two integrals are equal.

19. Prove that $\int_0^\infty \left(\frac{e^{-ax} - e^{-bx}}{x}\right)^2 dx = \log \frac{(2a)^{2a} (2b)^{2b}}{(a+b)^{2(a+b)}}$.

20. Shew that $\int_0^\infty \frac{\sqrt{x}\,.\log x}{(1+x)^2}\,dx = \pi$.

21. Shew that $\int_0^\infty \left(e^{-\frac{a^2}{x^2}} - e^{-\frac{b^2}{x^2}}\right) dx = (b-a)\sqrt{\pi}$.

(*Solutions of Senate-House Problems*, by O'Brien and Ellis, p. 44.)

22. Shew that $\int_0^\infty \log\frac{e^x+1}{e^x-1}\,dx = \frac{\pi^2}{4}$.

23. Prove that $\int_0^1 \frac{x^m - x^n}{\log x}\cdot\frac{dx}{x} = \log\frac{m}{n}$, and reconcile with this equation the result of transforming $\int_0^1 \frac{x^{r-1}\,dx}{\log x}$ by making $x^r = y$.

24. Shew that $\int_0^{\frac{\pi}{2}} \sin^n\theta\,d\theta = \frac{\sqrt{\pi}}{2}\cdot\frac{\Gamma\left(\frac{n+1}{2}\right)}{\Gamma\left(\frac{n+2}{2}\right)}$.

25. Shew that $\int_0^1 \frac{x^{l-1}(1-x)^{m-1}\,dx}{(b+cx)^{l+m}} = \frac{\Gamma(l)\,\Gamma(m)}{\Gamma(l+m)}\,\frac{1}{b^m(b+c)^l}$.

26. Shew that $\int_0^{\frac{\pi}{2}} \frac{\cos^{2l-1}\theta\,\sin^{2m-1}\theta\,d\theta}{(a\cos^2\theta + b\sin^2\theta)^{l+m}} = \frac{\Gamma(l)\,\Gamma(m)}{2\Gamma(l+m)}\,\frac{1}{a^l b^m}$.

27. Shew that $\int_0^{\frac{\pi}{2}} \frac{\tan^n\theta\,d\theta}{a\cos^2\theta + b\sin^2\theta} = \frac{\pi}{2\cos\frac{1}{2}n\pi}\,\frac{1}{a^{\frac{1-n}{2}}\,b^{\frac{1+n}{2}}}$,

n being less than unity.

28. Shew that $\int_0^\pi \frac{\sin^{n-1}\theta\,d\theta}{(\alpha+\beta\cos\theta)^n} = \frac{\left\{\Gamma\left(\frac{n}{2}\right)\right\}^2}{\Gamma(n)}\,\frac{2^{n-1}}{(\alpha^2-\beta^2)^{\frac{n}{2}}}$.

29. Shew that $\int_0^1 \frac{x^{m-1}\,dx}{(1-x^n)^{\frac{m}{n}}} = \frac{\pi}{n\sin\frac{m\pi}{n}}$.

30. Shew that $\int_0^1 \frac{x^{n-1}\,dx}{(1+cx)(1-x)^n} = \frac{\pi}{(1+c)^n \sin n\pi}$.

31. Shew that $\int_0^\infty \frac{\sin ax \sin^2 cx}{x}\,dx = 0,\ \pm\frac{\pi}{4},$ or $\pm\frac{\pi}{8}$, according to the values of a and c.

32. Trace the locus of the equation

$$y = \int_0^\infty \frac{\sin\theta \cos\theta x}{\theta}\,d\theta.$$

33. Trace the locus of the equation

$$\frac{y}{b} = \int_0^\pi \log\{1 - 2e^{-u}\cos\theta + e^{-2u}\}\,d\theta,$$

where $u = \sin\frac{x}{a}$.

34. Trace the locus of the equation

$$y = \int_{-\frac{\pi}{2}}^{\frac{\pi}{2}} \frac{x\cos\theta\,d\theta}{\sqrt{(x^2 + 2x\sin\theta + 1)}},$$

in which the sign of the square root is always taken so as to make the quantity in the denominator positive.

35. Shew that

$$\int_0^{\frac{\pi}{2}}\int_0^{\frac{\pi}{2}} \sin x \sin^{-1}(\sin x \sin y)\,dx\,dy = \frac{\pi^2}{4} - \frac{\pi}{2}.$$

36. Compare the results obtained from

$$\int_0^\infty\int_0^\infty \sin ax\, e^{-xy}\,dx\,dy,$$

by performing the integrations in different order.

37. Find the value of $\int_0^\infty e^{-\frac{x^2}{a^2}-\frac{b^2}{x^2}}\,dx$, and hence shew that

$$\int_0^\infty \left(\frac{x^2}{a^2}+\frac{a^2}{x^2}\right) e^{-\frac{x^2}{a^2}-\frac{a^2}{x^2}}\,dx = \frac{5a\sqrt{\pi}}{4e^2} = 5\int_0^\infty \left(\frac{x^2}{a^2}-\frac{a^2}{x^2}\right) e^{-\frac{x^2}{a^2}-\frac{a^2}{x^2}}\,dx.$$

38. Shew that

$$\iint \frac{\sqrt{(1-x^2-y^2)}}{\sqrt{(1+x^2+y^2)}}\,dx\,dy = \frac{\pi}{4}\left(\frac{\pi}{2}-1\right),$$

the integral being extended over all the positive values of x and y which make x^2+y^2 not greater than unity.

39. Shew that

$$\iiint \cdots \frac{dx\,dy\,dz\ldots}{\sqrt{(1-x^2-y^2-z^2-\ldots)}} = \frac{\pi^{\frac{n+1}{2}}}{2^n\,\Gamma\left\{\frac{n+1}{2}\right\}},$$

the number of variables being n, and the integration being extended over all positive values which make

$$x^2+y^2+z^2+\ldots\ldots$$

not greater than unity.

40. If $\qquad A_0+A_1x+A_2x^2+\ldots\ldots=F(x),$

and $\qquad a_0+a_1x+a_2x^2+\ldots\ldots=f(x),$

prove that $A_0a_0+A_1a_1x^2+A_2a_2x^4+\ldots\ldots$

$$=\frac{1}{2\pi}\int_0^\pi \{F(u)+F(v)\}\{f(u)+f(v)\}\,d\theta - A_0a_0,$$

where $\qquad u=xe^{\theta\sqrt{(-1)}}$ and $v=xe^{-\theta\sqrt{(-1)}}$.

41. If the sum of the series $a_0+a_1x+a_2x^2+\ldots\ldots$ can be expressed in a finite form, then the sum of the series $a_0^2+a_1^2x^2+a_2^2x^4+\ldots\ldots$ can be expressed by a definite integral. Prove this, and hence shew that the sum of the squares of the coefficients of the terms of the expan-

sion of $(1+x)^n$ when n is a positive whole number, may be expressed by

$$\frac{2^{2n+2}}{\pi}\int_0^{\frac{\pi}{2}} \cos^{2n}\theta \cos^2 n\theta \, d\theta - 1.$$

42. Prove that

$$\int_0^{\infty} \frac{\cos cx \, dx}{1+x^2} = \frac{\pi}{2}\left\{\frac{e^c}{1+0^{-c}} + \frac{e^{-c}}{1+0^c}\right\}.$$

43. Shew that

$$\int_0^{\frac{\pi}{2}} \phi(\sin 2x) \cos x \, dx = \int_0^{\frac{\pi}{2}} \phi(\cos^2 x) \cos x \, dx.$$

(Liouville's *Journal de Mathématiques*, Vol. XVIII. page 168.)

44. Shew that $$1 - \frac{x^2}{2^2} + \frac{x^4}{2^2 4^2} - \ldots\ldots$$

$$= \frac{2}{\pi}\int_0^{\frac{\pi}{2}} \cos(x \sin y) \, dy.$$

45. Prove that

$$\int_0^{\infty} x^{m-1} e^{-x^n} dx \int_0^{\infty} y^{n-m-1} e^{-y^n} dy = \frac{\pi}{n^2 \sin \frac{m\pi}{n}}.$$

(See Art. 66; and change the variable y to u where $y = ux$.)

46. Shew that

$$\int_{-\infty}^{\infty} e^{-(x^2 \cos 2\theta + \frac{a^2}{2x^2} \sin 2\theta)} \begin{matrix}\cos\\ \sin\end{matrix} \left\{x^2 \sin 2\theta + \frac{a^2}{2x^2} \cos 2\theta\right\} dx$$

$$= \pi^{\frac{1}{2}} e^{-a} \begin{matrix}\cos\\ \sin\end{matrix} (\theta + a);$$

θ being comprised between the limits $\pm \frac{\pi}{4}$.

CHAPTER XIII.

EXPANSION OF FUNCTIONS IN TRIGONOMETRICAL SERIES.

301. THE subject we are about to introduce is one of the most remarkable applications of the Integral Calculus, and although in an elementary work like the present, only an imperfect outline can be given of it, yet on account of the novelty of the methods, and the importance of the results, even such an outline may be of service to the student. For fuller information we may refer to the *Differential and Integral Calculus* of Professor De Morgan. The subject is also frequently considered in the writings of Poisson, for example, in his *Traité de Mécanique*, Vol. I. pp. 643—653; in his *Traité de la Chaleur;* and in different Memoirs in the *Journal de l'Ecole Polytechnique.* The student may also consult a Memoir by Professor Stokes, in the 8th Vol. of the *Cambridge Philosophical Transactions*, a Memoir by Sir W. Hamilton, in the 19th Vol. of the *Transactions of the Royal Irish Academy*, and a Memoir by Professor Boole, in the 21st Vol. of the same *Transactions.*

302. It is required to find the values of the m constants $A_1, A_2, A_3, \dots A_m$, so that the expression

$$A_1 \sin x + A_2 \sin 2x + A_3 \sin 3x + \dots\dots + A_m \sin mx$$

may coincide in value with an assigned function of x when x has the values θ, 2θ, $3\theta, \dots m\theta$, where $\theta = \dfrac{\pi}{m+1}$.

Let $f(x)$ denote the assigned function of x, then we have by hypothesis the following m equations from which the constants are to be determined,

$$f(\theta) = A_1 \sin\theta + A_2 \sin 2\theta + A_3 \sin 3\theta + \ldots\ldots + A_m \sin m\theta,$$

$$f(2\theta) = A_1 \sin 2\theta + A_2 \sin 4\theta + A_3 \sin 6\theta + \ldots\ldots + A_m \sin 2m\theta,$$

....

$$f(m\theta) = A_1 \sin m\theta + A_2 \sin 2m\theta + A_3 \sin 3m\theta + \ldots\ldots + A_m \sin mm\theta.$$

Multiply the first of these equations by $\sin r\theta$, the second by $\sin 2r\theta$, ..., the last by $\sin mr\theta$; then add the results. The coefficient of A_s on the second side will then be

$$\sin r\theta \sin s\theta + \sin 2r\theta \sin 2s\theta + \ldots\ldots + \sin mr\theta \sin ms\theta;$$

we shall now shew that this coefficient is zero if s be different from r, and equal to $\frac{1}{2}(m+1)$ when s is equal to r.

First suppose s different from r. Now twice the above coefficient is equal to the series

$$\cos(r-s)\theta + \cos 2(r-s)\theta + \ldots\ldots + \cos m(r-s)\theta,$$

diminished by the series

$$\cos(r+s)\theta + \cos 2(r+s)\theta + \ldots\ldots + \cos m(r+s)\theta.$$

The sum of the first series is known from Trigonometry to be equal to

$$\frac{\sin(2m+1)\dfrac{(r-s)\theta}{2} - \sin\dfrac{(r-s)\theta}{2}}{2\sin\dfrac{(r-s)\theta}{2}},$$

that is, to

$$\frac{\sin\left\{(r-s)\pi - \dfrac{(r-s)\theta}{2}\right\} - \sin\dfrac{(r-s)\theta}{2}}{2\sin\dfrac{(r-s)\theta}{2}}.$$

This expression vanishes when $r-s$ is an odd number, and is equal to -1 when $r-s$ is an even number.

The sum of the second series can be deduced from that of the first by changing the sign of s; hence this sum vanishes when $r+s$ is an odd number, and is equal to -1 when $r+s$ is an even number.

Thus when s is different from r, the coefficient of A_s is zero.

When s is equal to r, the coefficient becomes

$$\sin^2 r\theta + \sin^2 2r\theta + \ldots\ldots + \sin^2 mr\theta,$$

that is, $\dfrac{m}{2} - \frac{1}{2}\{\cos 2r\theta + \cos 4r\theta + \ldots\ldots + \cos 2mr\theta\}$.

And by the method already used it will be seen that the sum of the series of cosines is -1; thus the coefficient of A_r is $\frac{1}{2}(m+1)$.

Hence we obtain

$$A_r = \frac{2}{m+1}\{\sin r\theta f(\theta) + \sin 2r\theta f(2\theta) + \ldots\ldots + \sin mr\theta f(m\theta)\},$$

and thus by giving to r in succession the different integral values from 1 to m, the constants are determined.

Now suppose m to increase indefinitely, then we have ultimately

$$A_r = \frac{2}{\pi}\int_0^\pi \sin rv f(v)\, dv.$$

And as $f(x)$ now coincides in value with the expression

$$A_1 \sin x + A_2 \sin 2x + \ldots\ldots$$

for an infinite number of equidistant values of x between 0 and π, we may write the result thus

$$f(x) = \frac{2}{\pi}\Sigma_1^\infty \sin nx \int_0^\pi \sin nv f(v)\, dv,$$

where the symbol Σ_1^∞ indicates a summation to be obtained by giving to n every positive integral value.

303. The theorem and demonstration of the preceding article are due to Lagrange; we have given this demonstra-

tion partly because of its historical interest, and partly because it affords an instructive view of the subject. We shall however not stop to examine the demonstration closely, but proceed at once to the mode of investigation adopted by Poisson.

304. The following expansion may be obtained by ordinary Trigonometrical methods,

$$\frac{1-h^2}{1-2h\cos\frac{\pi(v-x)}{l}+h^2}=1+2h\cos\frac{\pi(v-x)}{l}$$
$$+2h^2\cos\frac{2\pi(v-x)}{l}+2h^3\cos\frac{3\pi(v-x)}{l}+\ldots,$$

h being less than unity, so that the series is convergent.

Multiply both sides by $\phi(v)$, and integrate with respect to v between the limits $-l$ and l; also make h approach to unity as its limit. On the right-hand side the different powers of h become ultimately unity. The numerator of the fraction on the left-hand side will ultimately vanish, and thus the integral would vanish *if the denominator of the fraction were never zero.* But *if x lies between $-l$ and l,* the term $\cos\frac{\pi(v-x)}{l}$ will become equal to unity during the integration, and thus the denominator of the fraction will be $(1-h)^2$, and will tend towards zero as h approaches unity. Thus the integral will not necessarily vanish; we proceed to ascertain its value. Let $v-x=z$ and $h=1-g$, thus

$$\int\frac{(1-h^2)\,\phi(v)\,dv}{1-2h\cos\frac{\pi(v-x)}{l}+h^2}=\int\frac{g(1+h)\,\phi(x+z)\,dz}{g^2+4h\sin^2\frac{\pi z}{2l}}.$$

Now the only part of the integral which has any sensible value, is that which arises from very small positive or negative values of z; thus we may put

$$\sin\frac{\pi z}{2l}=\frac{\pi z}{2l},$$

and $$\phi(x+z)=\phi(x);$$

and the integral becomes

$$g(1+h)\,\phi(x)\int\frac{dz}{g^2+\dfrac{h\pi^2z^2}{l^2}}=2g\phi(x)\int\frac{dz}{g^2+\dfrac{\pi^2z^2}{l^2}}$$

$$=\frac{2l\phi(x)}{\pi}\tan^{-1}\frac{\pi z}{gl}.$$

Suppose α and $-\beta$ to be the limits of z; we thus get

$$\frac{2l\phi(x)}{\pi}\left\{\tan^{-1}\frac{\pi\alpha}{gl}+\tan^{-1}\frac{\pi\beta}{gl}\right\}.$$

Hence, finally, when g is supposed to vanish, we have $2l\phi(x)$. Thus if x lies between $-l$ and l,

$$\phi(x)=\frac{1}{2l}\int_{-l}^{l}\phi(v)\,dv+\frac{1}{l}\,\Sigma_1^\infty\int_{-l}^{l}\phi(v)\cos\frac{n\pi(v-x)}{l}\,dv.$$

If however $x=l$ or $-l$, then the integral on the left-hand side has its sensible part when v is indefinitely near to l and $-l$; we should then have to perform the above process in both cases, but the integral with respect to z would only extend in the former case from $-\beta$ to 0, and in the latter from 0 to α. Hence instead of $2l\phi(l)$ on the left-hand side, we should have

$$l\phi(l)+l\phi(-l).$$

Thus we have determined the value of the right-hand member when x lies between l and $-l$, both inclusive; its value in other cases will be determined in Art. 311.

305. In the same way as the result in Art. 304 is found, we have, if we integrate between 0 and l,

$$\phi(x)=\frac{1}{2l}\int_0^l\phi(v)\,dv+\frac{1}{l}\,\Sigma_1^\infty\int_0^l\phi(v)\cos\frac{n\pi(v-x)}{l}\,dv\ldots\ldots(1);$$

this holds if x has any value between 0 and l; but when $x=0$ the left-hand member must be $\frac{1}{2}\phi(0)$, and when $x=l$ the left-hand member must be $\frac{1}{2}\phi(l)$. Thus we have determined the value of the right-hand member when x lies between l and $-l$, both inclusive; its value in other cases will be determined in Art. 311.

Similarly

$$0 = \frac{1}{2l}\int_0^l \phi(v)\,dv + \frac{1}{l}\Sigma_1^\infty \int_0^l \phi(v)\cos\frac{n\pi(v+x)}{l}\,dv \text{.......(2)};$$

this holds for any value of x between 0 and l; but when $x=0$ the left-hand member must be $\frac{1}{2}\phi(0)$, and when $x=l$ the left-hand member must be $\frac{1}{2}\phi(l)$.

From (1) and (2) by addition

$$\phi(x) = \frac{1}{l}\int_0^l \phi(v)\,dv + \frac{2}{l}\Sigma_1^\infty \cos\frac{n\pi x}{l}\int_0^l \cos\frac{n\pi v}{l}\phi(v)\,dv \text{...(3)}.$$

This holds for any value of x between 0 and l, both inclusive.

From (1) and (2) by subtraction

$$\phi(x) = \frac{2}{l}\Sigma_1^\infty \sin\frac{n\pi x}{l}\int_0^l \sin\frac{n\pi v}{l}\phi(v)\,dv \text{......(4)}.$$

This holds for any value of x between 0 and l both exclusive; and when $x=0$ or l, the left-hand member should be zero.

Equation (4) coincides with Lagrange's Formula.

We may observe that either of the formulæ (3) and (4) may be deduced from the other. Suppose we take (3) and write $\sin\frac{\pi x}{l}\phi(x)$ instead of $\phi(x)$. Thus

$$\sin\frac{\pi x}{l}\phi(x) = \frac{1}{l}\int_0^l \sin\frac{\pi v}{l}\phi(v)\,dv$$
$$+\frac{2}{l}\Sigma_1^\infty \cos\frac{n\pi x}{l}\int_0^l \cos\frac{n\pi v}{l}\sin\frac{\pi v}{l}\phi(v)\,dv.$$

Now $\cos\frac{n\pi v}{l}\sin\frac{\pi v}{l} = \frac{1}{2}\sin\frac{(n+1)\pi v}{l} - \frac{1}{2}\sin\frac{(n-1)\pi v}{l}$;

and therefore it will be found that the result may be exhibited thus,

$$\sin\frac{\pi x}{l}\phi(x) =$$

$$\frac{1}{l}\Sigma_1^\infty\left\{\cos\frac{(n-1)\pi x}{l} - \cos\frac{(n+1)\pi x}{l}\right\}\int_0^l \sin\frac{n\pi v}{l}\phi(v)\,dv;$$

also $$\cos\frac{(n-1)\pi x}{l} - \cos\frac{(n+1)\pi x}{l} = 2\sin\frac{n\pi x}{l}\sin\frac{\pi x}{l};$$

and therefore by division by $\sin\frac{\pi x}{l}$ we obtain the formula (4).

We will now give some examples.

306. Expand x in a series of sines. Take formula (4) of Art. 305, and suppose $l = \pi$; then

$$\int v\sin nv\,dv = -\frac{v\cos nv}{n} + \frac{\sin nv}{n^2};$$

therefore $\int_0^\pi v\sin nv\,dv = \frac{\pi}{n}$ if n be odd, and $-\frac{\pi}{n}$ if n be even.

Thus

$$x = 2\{\sin x - \tfrac{1}{2}\sin 2x + \tfrac{1}{3}\sin 3x - \tfrac{1}{4}\sin 4x + \ldots\ldots\}.$$

This holds for values of x between 0 and π, and as both sides vanish with x it holds when $x = 0$; and it is obvious that if it holds for any positive value of x it holds for the corresponding negative value; hence it holds for values of x between $-\pi$ and π, exclusive of these limiting values.

307. Expand $\cos x$ in a series of sines. Take formula (4) of Art. 305 and suppose $l = \pi$; then

$$\int\cos v\sin nv\,dv = \tfrac{1}{2}\int\{\sin(n+1)v + \sin(n-1)v\}\,dv$$

$$= -\tfrac{1}{2}\left\{\frac{\cos(n+1)v}{n+1} + \frac{\cos(n-1)v}{n-1}\right\};$$

therefore $$\int_0^\pi \cos v \sin nv \, dv = 0 \text{ if } n \text{ is odd,}$$

$$= \frac{2n}{n^2-1} \text{ if } n \text{ is even;}$$

therefore

$$\cos x = \frac{2}{\pi}\left\{\frac{4}{3}\sin 2x + \frac{8}{15}\sin 4x + \ldots + \frac{1+(-1)^n}{n^2-1} n \sin nx + \ldots\right\}.$$

This holds from $x = 0$ to $x = \pi$, exclusive of these limiting values.

308. Expand x in a series of cosines.

Take formula (3) of Art. 305, and suppose $l = \pi$; then

$$\int v \cos nv \, dv = \frac{v \sin nv}{n} + \frac{\cos nv}{n^2};$$

therefore $\int_0^\pi v \cos nv \, dv = 0$ if n be even, and $-\frac{2}{n^2}$ if n be odd; and

$$\int_0^\pi v \, dv = \frac{\pi^2}{2},$$

thus $$x = \frac{\pi}{2} - \frac{4}{\pi}\left\{\cos x + \frac{1}{3^2}\cos 3x + \frac{1}{5^2}\cos 5x + \ldots\ldots\right\}.$$

This holds from $x = 0$ to $x = \pi$ both inclusive.

If we put $x = \frac{\pi}{2} - y$, we obtain the following formula, which holds for any value of y between $-\frac{\pi}{2}$ and $\frac{\pi}{2}$, both inclusive,

$$y = \frac{4}{\pi}\left\{\sin y - \frac{1}{3^2}\sin 3y + \frac{1}{5^2}\sin 5y - \ldots\right\}.$$

309. Expand $e^{ax} - e^{-ax}$ in a series of sines.

Here $$\int_0^\pi (e^{av} - e^{-av}) \sin nv \, dv = -\frac{n(e^{a\pi} - e^{-a\pi})}{a^2 + n^2} \cos n\pi.$$

Therefore $\dfrac{\pi}{2}\dfrac{e^{ax}-e^{-ax}}{e^{a\pi}-e^{-a\pi}}=\dfrac{\sin x}{1^2+a^2}-\dfrac{2\sin 2x}{2^2+a^2}+\dfrac{3\sin 3x}{3^2+a^2}-\ldots\ldots$

310. Expand $e^{a(\pi-x)}+e^{-a(\pi-x)}$ in a series of cosines.

Here $\displaystyle\int_0^\pi \{e^{a(\pi-v)}+e^{-a(\pi-v)}\}\cos nv\,dv=\frac{a\,(e^{a\pi}-e^{-a\pi})}{a^2+n^2},$

and $$\int_0^\pi \{e^{a(\pi-v)}+e^{-a(\pi-v)}\}\,dv=\frac{e^{a\pi}-e^{-a\pi}}{a}.$$

Therefore $\dfrac{\pi}{2a}\dfrac{e^{a(\pi-x)}+e^{-a(\pi-x)}}{e^{a\pi}-e^{-a\pi}}=\dfrac{1}{2a^2}+\dfrac{\cos x}{1^2+a^2}+\dfrac{\cos 2x}{2^2+a^2}+\ldots\ldots$

311. We have shewn that the formula (3) of Art. 305 holds for any value of x between 0 and l both inclusive; it is easy to determine what the right-hand member is equal to when x lies beyond these limits. Suppose x positive, and between l and $2l$; put $x=2l-x'$ so that x' is less than l, then

$$\cos\frac{n\pi x}{l}=\cos\left(2n\pi-\frac{n\pi x'}{l}\right)=\cos\frac{n\pi x'}{l};$$

therefore the value of the right-hand member is $\phi(x')$. Next suppose x greater than $2l$; and suppose it equal to $2ml+x'$, where x' is less than $2l$; then

$$\cos\frac{n\pi x}{l}=\cos\frac{n\pi x'}{l},$$

so that the value is the same as it would be if x' were put instead of x; that is, the value is $\phi(x')$ if x' be less than l, and $\phi(2l-x')$ if x' be greater than l.

It is obvious that for any negative value of x the value is the same as for the corresponding positive value.

Similarly we may shew that if x is positive and $=2ml+x'$, the value of the right-hand side of equation (4) of Art. 305 is the same as if x' were put instead of x, and is $\phi(x')$ if x' be less than l, and $-\phi(2l-x')$ if x' be greater than l. And for negative values of x the value is the same numerically as for the corresponding positive value, but with an opposite sign.

312. It may be observed that in the fundamental demonstration of Art. 304, we suppose that when h approaches unity as a limit, the expression

$$\int h^n \phi(v) \cos \frac{n\pi(v-x)}{l}\, dv$$

may be replaced by

$$\int \phi(v) \cos \frac{n\pi(v-x)}{l}\, dv,$$

however large n may be. We may shew that no error arises from this supposition, by proving that the latter integral vanishes when n is increased indefinitely. We have

$$\int \phi(v) \cos \frac{n\pi(v-x)}{l}\, dv = \frac{l\phi(v)}{\pi n} \sin \frac{n\pi(v-x)}{l}$$
$$- \frac{l}{\pi n} \int \phi'(v) \sin \frac{n\pi(v-x)}{l}\, dv,$$

which shews that the integral on the left-hand side will vanish when n is infinite, at least if $\phi'(v)$ be not infinite.

313. We have not yet alluded to one of the most remarkable points in connexion with the formulæ (3) and (4) of Art. 305. In these formulæ $\phi(x)$ need not be a *continuous function;* for example, from $x=0$ to $x=a$ we might have $\phi(x)=f_1(x)$, then from $x=a$ to $x=b$ we might have $\phi(x)=f_2(x)$, then from $x=b$ to $x=c$ we might have $\phi(x)=f_3(x)$, then from $x=c$ to $x=l$ we might have $\phi(x)=f_4(x)$. The formula (3) for instance would still be true for all values of x between 0 and l inclusive, as is evident from the mode of demonstration, *except* for the values where the discontinuity occurs. When for example $x=a$, then the value of the right-hand member would not be $f_1(a)$ or $f_2(a)$ but $\frac{1}{2}\{f_1(a)+f_2(a)\}$. If therefore for $x=a$ we have $f_1(x)=f_2(x)$, the formula holds also when $x=a$.

314. Find an expression which shall be equal to c when x lies between 0 and a, and equal to zero when x lies between a and l.

Take formula (3) of Art. 305. Here $\phi(v) = c$ from $v=0$ to $v=a$, and then from $v=a$ to $v=l$ it is zero; thus

$$\int_0^l \cos\frac{n\pi v}{l}\,\phi(v)\,dv$$

becomes $$c\int_0^a \cos\frac{n\pi v}{l}\,dv = \frac{cl}{n\pi}\sin\frac{n\pi a}{l},$$

therefore the required expression is

$$\frac{ca}{l} + \frac{2c}{\pi}\left\{\sin\frac{\pi a}{l}\cos\frac{\pi x}{l} + \tfrac{1}{2}\sin\frac{2\pi a}{l}\cos\frac{2\pi x}{l} + \tfrac{1}{3}\sin\frac{3\pi a}{l}\cos\frac{3\pi x}{l} + \ldots\right\};$$

this will give $\frac{1}{2}c$ when $x = a$.

Or we may use formula (4) of Art. 305. Then

$$c\int_0^a \sin\frac{nv\pi}{l}\,dv = \frac{cl}{n\pi}\left(1 - \cos\frac{na\pi}{l}\right),$$

and we have for the required expression

$$\frac{2c}{\pi}\left\{\operatorname{vers}\frac{\pi a}{l}\sin\frac{\pi x}{l} + \tfrac{1}{2}\operatorname{vers}\frac{2\pi a}{l}\sin\frac{2\pi x}{l} + \tfrac{1}{3}\operatorname{vers}\frac{3\pi a}{l}\sin\frac{3\pi x}{l} + \ldots\ldots\right\};$$

this gives 0 when $x = 0$, and $\frac{1}{2}c$ when $x = a$.

315. Find an expression which shall be equal to kx from $x = 0$ to $x = \frac{l}{2}$, and equal to $k(l - x)$ from $x = \frac{l}{2}$ to $x = l$.

Here

$$\int_0^l \phi(v)\cos\frac{n\pi v}{l}\,dv = \int_0^{\frac{l}{2}} kv\cos\frac{n\pi v}{l}\,dv + \int_{\frac{l}{2}}^l k(l-v)\cos\frac{n\pi v}{l}\,dv$$

$$= \frac{kl^2}{\pi}\left\{\frac{1}{2n}\sin\frac{n\pi}{2} + \frac{1}{n^2\pi}\cos\frac{n\pi}{2} - \frac{1}{n^2\pi}\right\} + \frac{kl^2}{n\pi}\left(\sin n\pi - \sin\frac{n\pi}{2}\right)$$

$$- \frac{kl^2}{\pi}\left\{\frac{1}{n}\sin n\pi - \frac{1}{2n}\sin\frac{n\pi}{2} + \frac{\cos n\pi}{n^2\pi} - \frac{\cos\frac{n\pi}{2}}{n^2\pi}\right\}$$

$$= \frac{kl^2}{\pi^2 n^2}\left\{2\cos\frac{n\pi}{2} - \cos n\pi - 1\right\}.$$

This is $-\frac{4kl^2}{\pi^2 n^2}$ when n is of the form $4r+2$, and 0 in every other case, and

$$\int_0^l \phi(v)\,dv = k\int_0^{\frac{l}{2}} v\,dv + k\int_{\frac{l}{2}}^l (l-v)\,dv = \frac{kl^2}{4};$$

thus the required expression is

$$\frac{kl}{4} - \frac{8kl}{\pi^2}\left\{\frac{1}{2^2}\cos\frac{2\pi x}{l} + \frac{1}{6^2}\cos\frac{6\pi x}{l} + \ldots\right\}.$$

If we denote this by y, then from $x=0$ to $x=\frac{1}{2}l$ both inclusive $y=kx$, then from $x=\frac{1}{2}l$ to $x=l$ both inclusive $y=k(l-x)$; for values of x greater than l the values of y recur as shewn in Art. 311. Thus the value of y is the ordinate of the figure formed by measuring from the origin equal lengths along the axis of x to the right and left, and drawing on each base thus obtained the same isosceles triangle.

As another example we may propose the following: find a function $\phi(x)$ which shall be equal to x from $x=0$ to $x=\alpha$, then be equal to α from $x=\alpha$ to $x=\pi-\alpha$, and then be equal to $\pi - x$ from $x=\pi-\alpha$ to $x=\pi$.

The result is

$$\phi(x) = \frac{4}{\pi}\left\{\sin\alpha\sin x + \frac{1}{3^2}\sin 3\alpha\sin 3x + \frac{1}{5^2}\sin 5\alpha\sin 5x + \ldots\right\};$$

this is true from $x=0$ to $x=\pi$ both inclusive.

The student may verify the following examples.

If x be numerically less than a the expression

$$\frac{8a}{\pi^2}\Sigma_0^\infty\left\{\frac{\cos(2n+1)\frac{\pi x}{2a}}{2n+1}\right\}^2$$

is equal to $a-x$ if x be positive, and $a+x$ if x be negative.

Prove that for values of x between $-\pi$ and π inclusive

$$\frac{x^2}{4} = \frac{\pi^2}{12} - \cos x + \frac{\cos 2x}{2^2} - \frac{\cos 3x}{3^2} + \ldots\ldots$$

This may be obtained from Art. 308 by integration; or from equation (3) of Art. 305.

316. Other formulæ may be given analogous to those in Art. 305; we will here investigate some. We have from Art. 305

$$\phi(x) = \frac{1}{2l}\int_0^l \phi(v)\,dv + \frac{1}{l}\Sigma_1^\infty \int_0^l \phi(v)\cos\frac{n\pi(v-x)}{l}\,dv \ldots (1).$$

This holds when x has any value between 0 and l; but when $x=0$ the left-hand member must be $\frac{1}{2}\phi(0)$, and when $x=l$ the left-hand member must be $\frac{1}{2}\phi(l)$. In the same manner as this result was obtained we may also prove that

$$2\phi(x) = \frac{1}{2l}\int_0^l \phi(v)\,dv + \frac{1}{l}\Sigma_1^\infty \int_0^l \phi(v)\cos\frac{n\pi(v-x)}{2l}\,dv \ldots (2).$$

This holds when x has any value between 0 and l; but when $x=0$ the left-hand member must be $\phi(0)$, and when $x=l$ the left-hand member must be $\phi(l)$.

Subtract (1) from (2); thus

$$\phi(x) = \frac{1}{l}\Sigma_1^\infty \int_0^l \phi(v)\cos\frac{(2n-1)\pi(v-x)}{2l}\,dv \ldots\ldots (3).$$

This holds when x has any value between 0 and l; but when $x=0$ the left-hand member must be $\frac{1}{2}\phi(0)$, and when $x=l$ the left-hand member must be $\frac{1}{2}\phi(l)$.

Now in the same manner as (3) was obtained, we may obtain the following result, starting with $v+x$ instead of $v-x$,

$$0 = \frac{1}{l}\Sigma_1^\infty \int_0^l \phi(v)\cos\frac{(2n-1)\pi(v+x)}{2l}\,dv \ldots\ldots (4).$$

This holds when x has any value between 0 and l; but when $x=0$ the left-hand member must be $\frac{1}{2}\phi(0)$, and when $x=l$ the left-hand member must be $-\frac{1}{2}\phi(l)$.

From (3) and (4) by addition and subtraction we obtain

$$\phi(x) = \frac{2}{l}\Sigma_1^\infty \cos\frac{(2n-1)\pi x}{2l}\int_0^l \phi(v)\cos\frac{(2n-1)\pi v}{2l}\,dv \ldots (5),$$

$$\phi(x) = \frac{2}{l}\Sigma_1^\infty \sin\frac{(2n-1)\pi x}{2l}\int_0^l \phi(v)\sin\frac{(2n-1)\pi v}{2l}\,dv \ldots (6).$$

These hold when x has any value between 0 and l inclusive, except that when $x=0$ the left-hand member of (6) must be 0, and when $x=l$ the left-hand member of (5) must be 0.

317. We shall apply the formula (5) of the preceding article to establish a remarkable theorem first given by John Bernoulli. Let there be any curve AB the tangents of which at A and B are at right angles; let the involute of this curve be formed beginning at A, and denote it by AC; let the involute of AC be formed beginning at C; and so on continually; then the ultimate figure obtained will be a cycloid.

Let s be the length of the arc of the original curve measured from A to any point P; let ρ be the radius of curvature at P, and θ the inclination of the tangent at P to the tangent at A. Let ρ_1 be the radius of curvature at the corresponding point of the first involute, ρ_2 that of the second involute, ρ_3 that of the third involute; and so on. Then θ expresses the inclination of $\rho, \rho_2, \rho_4, \ldots$ to the normal of the original curve at A; and θ also expresses the inclination of $\rho_1, \rho_3, \rho_5, \ldots$ to the normal of the original curve at B. Moreover $\rho_1, \rho_3, \rho_5, \ldots$ vanish when $\theta=0$; and $\rho_2, \rho_4, \rho_6, \ldots$ vanish when $\theta=\frac{\pi}{2}$.

Now $\rho=\frac{ds}{d\theta}$, and $\rho_1=s$; thus $\rho_1=\int_0^\theta \rho\, d\theta$.

Similarly,
$$\rho_2=\int_\theta^{\frac{\pi}{2}} \rho_1\, d\theta,$$
$$\rho_3=\int_0^\theta \rho_2\, d\theta,$$
$$\rho_4=\int_\theta^{\frac{\pi}{2}} \rho_3\, d\theta,$$
and so on.

Now in formula (5) of the preceding article suppose $l=\frac{\pi}{2}$; then since ρ is some function of θ, we have
$$\rho=A_1 \cos\theta + A_3 \cos 3\theta + A_5 \cos 5\theta + \ldots$$

where A_1, A_3, A_5, ... are certain constants determined by that formula (5).

Thus

$$\rho_1 = A_1 \sin\theta + \frac{1}{3} A_3 \sin 3\theta + \frac{1}{5} A_5 \sin 5\theta + \ldots\ldots$$

$$\rho_2 = A_1 \cos\theta + \frac{1}{3^2} A_3 \cos 3\theta + \frac{1}{5^2} A_5 \cos 5\theta + \ldots\ldots$$

$$\rho_3 = A_1 \sin\theta + \frac{1}{3^3} A_3 \sin 3\theta + \frac{1}{5^3} A_5 \sin 5\theta + \ldots\ldots$$

....

Proceeding thus we obtain, when n is indefinitely large,

$$\rho_n = A_1 \sin\theta, \quad \text{or } \rho_n = A_1 \cos\theta;$$

and these equations represent a cycloid; see Art. 105.

We may proceed to examine the nature of the result when the tangents at the extremities of the original curve are not inclined at a right angle. Suppose these tangents to be inclined at an angle α; and put α for l in the formula (5) of the preceding article. Then we have

$$\rho = A_1 \cos\frac{\pi\theta}{2\alpha} + A_3 \cos\frac{3\pi\theta}{2\alpha} + A_5 \cos\frac{5\pi\theta}{2\alpha} + \ldots\ldots;$$

and by proceeding in the same way as before we arrive at the result

$$\rho_n = k \cos\frac{\pi\theta}{2\alpha}, \quad \text{or } \rho_n = k \sin\frac{\pi\theta}{2\alpha},$$

where $$k = A_1 \left(\frac{2\alpha}{\pi}\right)^n.$$

If k were a *finite* quantity, we should thus obtain an epicycloid if α is greater than $\frac{\pi}{2}$, and a hypocycloid in which the diameter of the revolving circle is less than the radius of the fixed circle if α is greater than $\frac{\pi}{2}$; see Arts. 110 and 111; and this is the usual statement of the results. But it will be observed that k becomes indefinitely great in the former case and indefinitely small in the latter case; so that in the former case the radii of the fixed and revolving circles must be supposed to increase indefinitely, and in the latter case to diminish indefinitely.

318. In the formula

$$\phi(x) = \frac{1}{2l}\int_{-l}^{l} \phi(v)\,dv + \frac{1}{l}\Sigma_1^\infty \int_{-l}^{l} \cos\frac{n\pi(v-x)}{l}\,\phi(v)\,dv,$$

suppose l to increase without limit; then if $\phi(v)$ be such that the term $\frac{1}{2l}\int_{-l}^{l} \phi(v)\,dv$ vanishes with $\frac{1}{l}$ we have

$$\phi(x) = \frac{1}{\pi}\int_0^\infty \int_{-\infty}^\infty \cos u(v-x)\,\phi(v)\,du\,dv.$$

This is called Fourier's Theorem.

MISCELLANEOUS EXAMPLES.

1. Change the order of integration in the expression

$$\int_0^a \int_{\frac{a^2-x^2}{2a}}^{\sqrt{(a^2-x^2)}} \phi(x, y)\,dx\,dy.$$

2. Change the order of integration in the expression

$$\int_0^{2a} \int_{\sqrt{(2ax-x^2)}}^{\sqrt{(4ax)}} \phi(x, y)\,dx\,dy.$$

3. Transform $\int_0^c \int_{ax}^{bx} \phi(x, y)\,dx\,dy$ into an integral with respect to u and v, having given $u = y + x$, $y = uv$; and determine the limits of the new integral.

4. Transform $\int_0^a \int_0^b \phi(x, y)\,dx\,dy$ into an integral with respect to u and v, having given $y + cx = u$, $y = uv$; and determine the limits of the new integral.

5. Transform the integral

$$\iiint (x-y)(y-z)(z-x)\,dx\,dy\,dz$$

into one in which u, v, w are the independent variables, where

$$u^3 = xyz, \quad \frac{1}{v} = \frac{1}{x} + \frac{1}{y} + \frac{1}{z}, \quad w^2 = x^2 + y^2 + z^2.$$

6. Prove that

$$\left\{\int_0^\infty e^{-\tau}\,dx\right\}^2 = \frac{1}{2}\int_0^\infty e^{-t}\,dx \int_0^\infty \frac{dx}{(1+x^{2n})^{\frac{1}{n}}},$$

where $t = x^n$ and $\tau = t^2$.

(See Arts. 263 and 66; and transform as in Art. 242.)

7. Prove by transforming the expression from rectangular to polar co-ordinates that the value of the definite integral

$$\int_0^\infty \int_0^\infty e^{-(x^4+2x^2y^2\cos\alpha+y^4)}\,dx\,dy$$

is equal to $\frac{1}{4}\sqrt{\pi}\,F\left(\sin\frac{\alpha}{2}\right)$, where $F\left(\sin\frac{\alpha}{2}\right)$ denotes a complete elliptic function of the first order of which $\sin\frac{\alpha}{2}$ is the modulus.

8. Prove that $\int_0^{\frac{\pi}{4}} \tan\theta \log\cot\theta\,d\theta = \frac{\pi^2}{48}$.

9. Prove that

$$\int_0^\infty e^{-x^2 n\cot 2\beta} \sin(nx^2+\alpha)\,dx = \sin(\alpha+\beta)\sqrt{\left(\frac{\pi\sin 2\beta}{4n}\right)}.$$

10. Shew that

$$\int_0^{\frac{\pi}{4}} \tan^{-1}\{n\sqrt{(1-\tan^2 x)}\}\,dx = \frac{\pi}{2}\tan^{-1} n\sqrt{2} - \frac{\pi}{2}\cot^{-1}\frac{\sqrt{(1+n^2)}}{n}.$$

11. If $f(\xi) = \int_0^\pi \frac{\sin\left(\xi \tan \frac{\theta}{2}\right)}{\sin \theta} d\theta$, determine the geometrical meaning of the equation $y = xf(\sin x)$.

12. A curve of double curvature revolves round the axis of x; shew that the surface generated

$$= 2\pi \int \sqrt{\{(ydy + zdz)^2 + (y^2 + z^2)(dx)^2\}}.$$

CHAPTER XIV.

APPLICATION OF THE INTEGRAL CALCULUS TO QUESTIONS OF MEAN VALUE AND PROBABILITY.

319. We will here give a few simple examples of the application of the Integral Calculus to questions relating to *mean value* and to *probability*.

Let $\phi(x)$ denote any function of x, and suppose x successively equal to a, $a+h$, $a+2h$, ... $a+(n-1)h$. Then

$$\frac{\phi(a)+\phi(a+h)+\phi(a+2h)+\ldots+\phi\{a+(n-1)h\}}{n}$$

may be said to be the *mean* or *average* of the n values which $\phi(x)$ receives corresponding to the n values of x. Let

$$b-a=nh,$$

then the above *mean value* may be written thus,

$$\frac{[\phi(a)+\phi(a+h)+\phi(a+2h)+\ldots+\phi\{a+(n-1)h\}]\,h}{b-a}.$$

Suppose a and b to remain fixed and n to increase indefinitely; then the limit of the above expression is

$$\frac{\int_a^b \phi(x)\,dx}{b-a}.$$

This may accordingly be defined to be the *mean value* of $\phi(x)$ when x varies continuously between a and b.

320. As an example we may take the following question; find the mean distance of all points within a circle from a fixed point on the circumference. By this enunciation we intend the following process to be performed. Let the area of a circle be divided into a large number n of equal small areas; form a fraction of which the numerator is the sum of the distances of these small areas from a fixed point on the circumference, and the denominator is n; then find the limit of this fraction when n is infinite.

Suppose $r_1, r_2, \ldots r_n$ to denote the respective distances of the small areas; then the fraction required is

$$\frac{1}{n}\{r_1 + r_2 + \ldots + r_n\}.$$

Multiply both numerator and denominator by $r\,\Delta\theta\,\Delta r$, which represents the area of a small element (Art. 148), thus the fraction becomes

$$\frac{\{r_1 + r_2 + \ldots + r_n\}\, r\,\Delta\theta\,\Delta r}{nr\,\Delta\theta\,\Delta r}.$$

The limit of the denominator will represent the area of the circle, that is, πc^2, if c be the radius of the circle. The limit of the numerator will be, by the definitions of the Integral Calculus, $\iint r^2\, d\theta\, dr$, the limits being so taken as to include all the elements of area within the boundary of the circle. Thus the result is

$$\frac{\int_{-\frac{\pi}{2}}^{\frac{\pi}{2}} \int_0^{2c\cos\theta} r^2\, d\theta\, dr}{\pi c^2}.$$

This will be found to give $\dfrac{32c}{9\pi}$.

321. The equation to a curve is $r = c \sin\theta \cos\theta$, find the *mean length* of all the radii vectores drawn at equal angular intervals in the first quadrant.

It easily follows, as in the last article, that the required *mean length* is

$$\frac{\int_0^{\frac{\pi}{2}} c \sin\theta \cos\theta \, d\theta}{\frac{\pi}{2}},$$

that is, $\frac{c}{\pi}$.

Again, suppose the portion of this curve which lies in the first quadrant to revolve round the initial line, and thus to generate a surface. Let radii vectores be drawn from the origin to different points of the surface *equably in all directions:* it is required to find the mean length of the radii vectores.

The only difficulty in this question lies in apprehending clearly what is meant by the words in Italics. Conceive a spherical surface having the origin as centre; then by equable angular distribution of the radii vectores, we mean that they are to be so drawn that the number of them which fall upon any portion of the spherical surface must be proportional to the area of that portion. Now the area of any portion of a sphere of radius a is found by integrating

$$a^2 \iint \sin\theta \, d\phi \, d\theta$$

within proper limits (Art. 175). Hence $a^2 \sin\theta \, \Delta\phi \, \Delta\theta$ may be taken to denote an element of a spherical surface, and $2\pi a^2$ is the area of half the surface of a sphere. Thus we shall have as the required result

$$\frac{\iint a^2 c \sin\theta \cos\theta \sin\theta \, d\phi \, d\theta}{2\pi a^2},$$

the limits being so taken as to extend the integrations over the entire surface considered.

Hence we obtain

$$\frac{\int_0^{2\pi}\int_0^{\frac{\pi}{2}} c \sin^2\theta \cos\theta \, d\phi \, d\theta}{2\pi},$$

that is, $\frac{c}{3}$.

322. A large plane area is ruled with parallel equidistant lines; a thin rod, the length of which is less than the distance between two consecutive lines, is thrown at hazard on the area; find the probability that the rod will fall across one of the lines.

Let $2a$ be the distance between two consecutive lines and $2c$ the length of the rod. It is easily seen that we do not alter the problem by supposing the centre of the rod constrained to fall upon a line drawn between consecutive lines of the given system and meeting them at right angles, for the proportion of the favourable cases to the whole number of cases remains the same after this limitation as before.

Let the centre of the rod be at a distance x from the nearer of the two selected parallels; then suppose the rod to revolve round its centre, and it is obvious that in this position of its centre the chance that it crosses the line is $\frac{4\phi}{2\pi}$, where

$$c \cos \phi = x.$$

And we may denote by $\frac{\Delta x}{a}$ the chance that the centre of the rod falls between the distances x and $x + \Delta x$ from the nearer of the two parallels. Thus the chance required will be denoted by the limit of the sum of such quantities as $\frac{2\phi}{\pi} \frac{\Delta x}{a}$ that is, it will be

$$\frac{2}{\pi a}\int \phi \, dx,$$

where $\cos\phi = \frac{x}{c}$.

The limits of x are 0 and c; hence the result

$$= \frac{2c}{\pi a}\int_0^{\frac{\pi}{2}} \phi \sin\phi \, d\phi$$

$$= \frac{2c}{\pi a}.$$

EXAMPLES.

1. If $r = f(\theta)$ and $y = f\left(\frac{x}{a}\right)$ be the equations to two curves, $f(\theta)$ being a function which vanishes for the values θ_1, θ_2, and is positive for all values between these limits, and if A be the area of the former between the limits $\theta = \theta_1$ and $\theta = \theta_2$, and M the arithmetical mean of all the transverse sections of the solid generated by the revolution about the axis of x of the portion of the latter curve between the limits $x = a\theta_1$ and $x = a\theta_2$, shew that

$$M = \frac{2\pi}{\theta_2 - \theta_1} A,$$

supposing θ_2 greater than θ_1.

2. A ball is fired at random from a gun which is equally likely to be presented in any direction in space above the horizon; shew that the chance of its reaching more than $\frac{1}{m}$th of its greatest range is $\sqrt{\left(1 - \frac{1}{m}\right)}$.

3. From a point in the circumference of a circular field a projectile is thrown at random with a given velocity, which is such that the diameter of the field is equal to the greatest range of the projectile; find the chance of its falling within the field.

Result. $\frac{1}{2} - \frac{2}{\pi}(\sqrt{2} - 1)$.

4. On a table a series of straight lines at equal distances from one another is drawn, and a cube is thrown at random on the table. Supposing the diagonal of the cube less than the distance between consecutive lines, find the chance that the cube will rest without covering any part of the lines.

Result. $\dfrac{4a}{c\pi}$, where a is the edge of the cube and c the distance of consecutive lines.

5. Prove that the mean of all the radius-vectors of an ellipse, the focus being the origin, is equal to half the minor axis, when the lines are drawn at equal angular intervals; and is equal to half the major axis when the lines are drawn so that the abscissæ of their extremities increase uniformly.

6. An indefinite number of equidistant parallel lines are drawn on a plane, and a rod whose length is equal to r times the perpendicular distance between two consecutive lines is thrown at random on the plane; find the probability of its falling upon n of the lines. If $n = r = 1$, shew that the probability is $\dfrac{2}{\pi}$.

7. Two arrows are sticking in a circular target; what is the chance that their distance is greater than the radius of the target?

Result. $\dfrac{3\sqrt{3}}{4\pi}$.

8. Supposing the orbits of comets to be equally distributed through space, prove that their mean inclination to the plane of the ecliptic is the angle subtended by an arc equal to the radius.

9. A certain territory is bounded by two meridian circles and by two parallels of latitude which differ in longitude and latitude respectively by one degree, and is known to lie within certain limits of latitude; find the probable superficial area.

10. A line is taken of given length a, and two other lines are taken each less than the first line and laid down in it at hazard, any one position of either being as likely as any other. The lengths of these lines are b and b'; it is required to find the probability that they shall not have a part exceeding c in common.

Result. $\dfrac{(a-b-b'+c)^2}{(a-b)(a-b')}$.

Camb. Phil. Transactions, Vol. VIII. p. 386.

11. An indefinitely large plane area is ruled with parallel equidistant lines, the distance between consecutive lines being c. A closed curve which has no singular points whose greatest diameter is less than c is thrown down on the area. Shew that the chance that the curve falls on one of the lines is $\dfrac{l}{\pi c}$, where l denotes the perimeter of the curve.

12. A messenger M starts from A towards B (distance a) at a rate of v miles per hour, but before he arrives at B a shower of rain commences at A and at all places occupying a certain distance z towards, but not reaching beyond, B, and moves at the rate of u miles an hour towards A; if M be caught in this shower he will be obliged to stop until it is over; he is also to receive for his errand a number of shillings inversely proportional to the time occupied in it, at the rate of n shillings for one hour. Supposing the distance z to be unknown, as also the time at which the shower commenced, but all events to be equally probable, shew that the value of M's expectation is, in shillings,

$$\frac{nv}{a}\left\{\frac{1}{2}-\frac{u}{v}+\frac{u(u+v)}{v^2}\log\frac{u+v}{u}\right\}.$$

CHAPTER XV.

CALCULUS OF VARIATIONS.

Maxima and Minima of integrals involving one dependent variable with fixed limits.

323. THE theory of maxima and minima values of given functions is fully considered in works on the Differential Calculus. If, for example, y denotes any given function of an independent variable x, then we can find the value or values of x which make y a maximum or minimum, or we can shew that there are no such values in some cases.

We are now however about to consider a new class of maxima and minima problems. Let y denote a function of x which is at present undetermined; and let V denote a given function of $x, y, \frac{dy}{dx}, \frac{d^2y}{dx^2}, \dots$ Suppose we wish to find the relation which must hold between x and y in order that the integral $\int Vdx$, taken between given limits, may have a maximum or minimum value. We cannot here effect the integration because y is not known as a function of x, and therefore V is not known as a function of x; thus the ordinary methods of solving maxima and minima problems do not apply. We require then a new method, which we shall now proceed to explain.

324. The department of analysis to which we are about to introduce the student is called the *Calculus of Variations*; its object is to find the maxima or minima values of integral expressions, the expressions being supposed to vary by

assigning different *forms* to the functions denoted by the dependent variables. It will be seen, as we proceed, that the method of finding these maxima or minima values is analogous to that of finding ordinary maxima or minima values by the Differential Calculus.

325. It will be useful to recur to the method given in the Differential Calculus. The student will remember that the terms *maximum* and *minimum* are technical terms, which are defined and illustrated in treatises on the Differential Calculus; and they are used in mathematics in the sense there assigned to them. Mistakes are frequently made by confounding *a maximum value* in the technical sense of the word maximum, with *the greatest value* in the ordinary sense of the word greatest.

Suppose y a given function of an independent variable x; then if an indefinitely small change is given to x, in general an indefinitely small change is consequently given to y, which is comparable in magnitude with that given to x. The process of finding a maximum or minimum value of y may be said to consist of two parts. First we determine such a value of x that an indefinitely small change in it does not produce in y a comparable indefinitely small change, but a change which is indefinitely small compared with that of x. In the second place, we examine the sign of this indefinitely small change which is produced in y by the change of x; and for a maximum this sign is to be necessarily negative, and for a minimum positive.

We may therefore describe this process briefly thus; we make the terms of the first order in the change of the dependent variable vanish, and we examine the sign of the terms of the second order. We shall pursue a similar method with the problem which we have now to discuss; we confine ourselves, however, at present entirely to the first part of the process, and shall hereafter recur to the second part.

326. We have first to explain the notation which will be used. Let x denote an independent variable, y any function of x, and $\frac{dy}{dx}, \frac{d^2y}{dx^2}, \ldots$ the differential coefficients of y

with respect to x. We shall use δy to denote an indefinitely small quantity which may be any function of x; and if u denote any quantity whatever which depends on y we shall denote by δu the increment which u receives when y is changed into $y + \delta y$. Thus, for example, consider the differential coefficient $\frac{dy}{dx}$; when y receives the increment δy this differential coefficient receives the increment $\frac{d\delta y}{dx}$, so that by $\delta \frac{dy}{dx}$ we mean $\frac{d\delta y}{dx}$. It is often convenient to use the symbol p for $\frac{dy}{dx}$; and so also δp is a convenient symbol for $\frac{d\delta y}{dx}$. Again, consider the second differential coefficient $\frac{d^2y}{dx^2}$; when y receives the increment δy this second differential coefficient receives the increment $\frac{d^2\delta y}{dx^2}$, and as the second differential coefficient is often denoted by q we may conveniently use δq for $\frac{d^2\delta y}{dx^2}$. Similarly r and s may be used for the third and fourth differential coefficients of y respectively, and δr and δs for $\frac{d^3\delta y}{dx^3}$ and $\frac{d^4\delta y}{dx^4}$ respectively; and so on.

The differential coefficients are also often denoted by $y', y'', y''', \ldots$; and thus $\delta y', \delta y'', \delta y''', \ldots$ may be used as equivalent to $\delta p, \delta q, \delta r, \ldots$ respectively.

327. The introduction of the symbol δ is due to Lagrange. The student will see that this symbol resembles in meaning the symbol d, which is used in the Differential Calculus. Both dy and δy express indefinitely small increments; dy however is generally used to denote the change in *value* of a *given* function consequent upon a change in the value of the dependent variable, δy is used to denote the change made by ascribing an arbitrary change to the *form* of a function. The quantity denoted by δy is called the *variation* of y.

328. Let V denote a given function of x, y, $\frac{dy}{dx}$, $\frac{d^2y}{dx^2}$, ...; and let $U=\int_{x_0}^{x_1} V\,dx$, where x_0 and x_1 are supposed to denote given limits. The value of U cannot be found so long as we do not know what particular function y is of x; but without knowing this we are able to obtain an expression for the increment made in U by ascribing the arbitrary increment δy to y, from which important inferences can be drawn.

Suppose $$V=\phi(x, y, y', y'', y''', \ldots);$$
then by definition
$$\delta V=\phi(x, y+\delta y,\ y'+\delta y',\ y''+\delta y'',\ y'''+\delta y''', \ldots) - \phi(x, y, y', y'', y''', \ldots).$$

The first term may be expanded by the ordinary extension of Taylor's theorem; thus

$$\delta V=\frac{dV}{dy}\delta y+\frac{dV}{dy'}\delta y'+\frac{dV}{dy''}\delta y''+\frac{dV}{dy'''}\delta y'''+\ldots,$$

where $\frac{dV}{dy}$ is the partial differential coefficient of V with respect to y, also $\frac{dV}{dy'}$ is the partial differential coefficient of V with respect to y'; and so on.

In the above expression for δV we have only expressed *terms of the first order*, that is, we have omitted the terms of the second and higher orders with respect to the small quantities δy, $\delta y'$,.... This we shall continue to do throughout the remainder of the investigation.

Then

$$\delta U=\int_{x_0}^{x_1}\delta V\,dx$$
$$=\int_{x_0}^{x_1}\left\{\frac{dV}{dy}\delta y+\frac{dV}{dy'}\delta y'+\frac{dV}{dy''}\delta y''+\frac{dV}{dy'''}\delta y'''+\ldots\right\}dx.$$

We shall now transform this expression by integration by parts. For shortness put

$$\frac{dV}{dy} = N, \quad \frac{dV}{dy'} = P, \quad \frac{dV}{dy''} = Q, \quad \frac{dV}{dy'''} = R, \ldots$$

Then $$\int P\delta y' dx = \int P \frac{d\delta y}{dx} dx = P\delta y - \int \frac{dP}{dx} \delta y \, dx;$$

therefore $$\int_{x_0}^{x_1} P\delta y' dx = (P\delta y)_1 - (P\delta y)_0 - \int_{x_0}^{x_1} \frac{dP}{dx} \delta y \, dx.$$

Here $(P\delta y)_1$ is used to denote the value of $P\delta y$ when x_1 is put for x, and $(P\delta y)_0$ is used to denote the value of $P\delta y$ when x_0 is put for x; a similar notation will be used throughout. It is to be carefully observed that $\frac{dP}{dx}$ means the *complete* differential coefficient of P with respect to x, that is to say, in forming $\frac{dP}{dx}$ we are to remember that y and its differential coefficients all involve x implicitly.

Again

$$\int Q\delta y'' dx = \int Q \frac{d^2\delta y}{dx^2} dx = Q\frac{d\delta y}{dx} - \int \frac{dQ}{dx}\frac{d\delta y}{dx} dx$$

$$= Q\frac{d\delta y}{dx} - \frac{dQ}{dx}\delta y + \int \frac{d^2 Q}{dx^2} \delta y \, dx;$$

therefore

$$\int_{x_0}^{x_1} Q\delta y'' dx = \left(Q\frac{d\delta y}{dx} - \frac{dQ}{dx}\delta y\right)_1 - \left(Q\frac{d\delta y}{dx} - \frac{dQ}{dx}\delta y\right)_0 + \int_{x_0}^{x_1} \frac{d^2 Q}{dx^2} \delta y \, dx.$$

Similarly

$$\int_{x_0}^{x_1} R\delta y''' dx = \left(R\frac{d^2\delta y}{dx^2} - \frac{dR}{dx}\frac{d\delta y}{dx} + \frac{d^2 R}{dx^2}\delta y\right)_1$$

$$- \left(R\frac{d^2\delta y}{dx^2} - \frac{dR}{dx}\frac{d\delta y}{dx} + \frac{d^2 R}{dx^2}\delta y\right)_0 - \int_{x_0}^{x_1} \frac{d^3 R}{dx^3} \delta y \, dx.$$

This process may be continued until all the symbols $\delta y'$, $\delta y''$, $\delta y'''$, $\delta y''''$, ... are brought from under the integral sign. It is to be observed that all the differential coefficients $\frac{dQ}{dx}$, $\frac{d^2Q}{dx^2}$, $\frac{dR}{dx}$, $\frac{d^2R}{dx^2}$, $\frac{d^3R}{dx^3}$ are *complete* differential coefficients.

Hence finally

$$\delta U = \delta y_1 \left\{P - \frac{dQ}{dx} + \frac{d^2R}{dx^2} - \dots\right\}_1 - \delta y_0 \left\{P - \frac{dQ}{dx} + \frac{d^2R}{dx^2} - \dots\right\}_0$$

$$+ \delta p_1 \left\{Q - \frac{dR}{dx} + \dots\right\}_1 - \delta p_0 \left\{Q - \frac{dR}{dx} + \dots\right\}_0$$

$$+ \delta q_1 \{R - \dots\}_1 - \delta q_0 \{R - \dots\}_0$$

$$+ \dots\dots\dots\dots\dots\dots$$

$$+ \int_{x_0}^{x_1} \left(N - \frac{dP}{dx} + \frac{d^2Q}{dx^2} - \frac{d^3R}{dx} + \dots\right) \delta y \, dx.$$

Here we have adopted some obvious simplifications of notation; thus we use δy_1 for $(\delta y)_1$, and δp_1 for $\left(\frac{d\delta y}{dx}\right)_1$, and so on.

329. The value of δU may be denoted thus,

$$\delta U = H_1 - H_0 + \int_{x_0}^{x_1} K \delta y \, dx,$$

where H_1 denotes a certain aggregate of terms in which x_1 is put for x, and H_0 a similar aggregate of terms in which x_0 is put for x; these aggregates do not involve any integrations. Also

$$K = N - \frac{dP}{dx} + \frac{d^2Q}{dx^2} - \frac{d^3R}{dx^3} + \dots$$

Since $H_1 - H_0$ involves only the values of the variables at the *limits*, we shall sometimes speak of $H_1 - H_0$ as the *terms at the limits*.

330. We can now determine the conditions that must hold in order that U may have a maximum or minimum value. For, in order that U may have a maximum or minimum value, δU must vanish, whatever δy may be, provided only that it is an indefinitely small quantity. This requires that

$$K = 0 \text{ and } H_1 - H_0 = 0.$$

For if K is not always zero, it will be in our power to give such a value to δy as will make δU positive or negative at our pleasure, and not zero. Suppose, for example, that the highest differential coefficient of δy which occurs in $H_1 - H_0$ is the nth. Put $\delta y = \alpha(x - x_1)^{2n}(x - x_0)^{2n}$, where α is a function of x which is indefinitely small, and is at present undetermined. Then this value of δy makes $H_1 - H_0$ vanish, so that δU reduces to $\int_{x_0}^{x_1} K\delta y\, dx$. Now take α such that it is always positive when K is positive, and negative when K is negative; then δU is necessarily positive. And if the sign of α be changed, δU is necessarily negative. Thus if K is not always zero, it is in our power so to take δy as to make δU positive or negative at our pleasure.

Hence for a maximum or minimum value of U we must have $K = 0$; and then $\int_{x_0}^{x_1} K\,\delta y\, dx$ vanishes, and therefore also $H_1 - H_0$ must $= 0$.

331. The student has now become acquainted with the essential features of the Calculus of Variations; these are (1) the reduction of δU to the form $H_1 - H_0 + \int_{x_0}^{x_1} K\,\delta y\, dx$, (2) the principle that K must vanish in order that U may be a maximum or minimum. Although the subject admits of considerable development, by various extensions of the problem we have considered, still the two results we have already obtained are the chief results.

332. We now proceed to examine more closely the nature of the two conditions

$$K = 0 \text{ and } H_1 - H_0 = 0.$$

The equation $K = 0$ is what is called a *differential equation*. Suppose that $\frac{d^3y}{dx^3}$ is the highest differential coefficient which occurs in V; then this will in general occur in R also, and therefore in $\frac{d^3R}{dx^3}$ the differential coefficient $\frac{d^6y}{dx^6}$ will occur, and this will be the highest differential coefficient which occurs in K, so that the differential equation $K = 0$ will be of the sixth order. And in general the order of the differential equation is twice the order of the highest differential coefficient which occurs in V.

It is shewn in treatises on Differential Equations that the solution of a differential equation involves as many arbitrary constants as the number which expresses the order of the differential equation. We must now shew how the arbitrary constants which arise from the solution of the equation $K = 0$ are to be determined, so that a definite result may be obtained. The condition $H_1 - H_0 = 0$ serves for this purpose. Two cases may arise.

(1) Suppose that no conditions are imposed by the problem on the values of y and its differential coefficients at the limits of the integration; then δy_1, δy_0, δp_1, δp_0,... are all arbitrary quantities, that is, we have it in our power to suppose any indefinitely small values we please for these quantities; for example, we may suppose that as many of them as we please are zero. Since δy_1, δy_0, δp_1, δp_0,... are thus all arbitrary, in order that $H_1 - H_0$ may certainly vanish, the coefficient of each of the arbitrary quantities must vanish. This furnishes for determining the constants as many equations as there are constants.

(2) Suppose that conditions are imposed by the problem upon the values of y and its differential coefficients at the limits of the integration; then δy_1, δy_0, δp_1, δp_0,... are not *all* arbitrary, for some of them can be expressed in terms of the

rest by means of the given conditions. Let as many as possible of the quantities δy_1, δy_0, δp_1, δp_0,... be eliminated from $H_1 - H_0$, and then the coefficients of those which remain must be equated to zero. The equations thus obtained, together with those which express the given conditions, will form a system equal in number to the number of constants, and therefore will serve to determine those constants.

333. The principal difficulty in examples consists in the solution of the differential equation $K=0$, and this difficulty is frequently insuperable.

We will now shew that when V does not explicitly contain the independent variable, one step in the solution of the differential equation can always be taken. It will be sufficient for practical purposes to confine ourselves to the case in which V involves no differential coefficient of y higher than the third.

Since V is supposed not to involve x explicitly, we have for the complete differential coefficient of V

$$\frac{dV}{dx} = N\frac{dy}{dx} + P\frac{dp}{dx} + Q\frac{dq}{dx} + R\frac{dr}{dx}.$$

And by supposition

$$0 = N - \frac{dP}{dx} + \frac{d^2Q}{dx^2} - \frac{d^3R}{dx^3} \quad \text{.............. (1).}$$

Thus

$$\frac{dV}{dx} = \frac{dP}{dx}\frac{dy}{dx} + P\frac{dp}{dx} - \frac{d^2Q}{dx^2}\frac{dy}{dx} + Q\frac{dq}{dx} + \frac{d^3R}{dx^3}\frac{dy}{dx} + R\frac{dr}{dx}.$$

Now

$$\frac{dP}{dx}\frac{dy}{dx} + P\frac{dp}{dx} = \frac{d}{dx}P\frac{dy}{dx},$$

$$\frac{d^2Q}{dx^2}\frac{dy}{dx} - Q\frac{dq}{dx} = \frac{d}{dx}\left\{\frac{dQ}{dx}\frac{dy}{dx} - Q\frac{d^2y}{dx^2}\right\},$$

$$\frac{d^3R}{dx^3}\frac{dy}{dx} + R\frac{dr}{dx} = \frac{d}{dx}\left\{\frac{d^2R}{dx^2}\frac{dy}{dx} - \frac{dR}{dx}\frac{d^2y}{dx^2} + R\frac{d^3y}{dx^3}\right\}.$$

Hence, by integration,

$$V = P\frac{dy}{dx} + \frac{dQ}{dx}\frac{dy}{dx} - Q\frac{d^2y}{dx^2} + \frac{d^2R}{dx^2}\frac{dy}{dx} - \frac{dR}{dx}\frac{d^2y}{dx^2} + R\frac{d^3y}{dx^3} + C \quad (2),$$

where C is an arbitrary constant. The highest differential coefficient that can occur in (2) is $\frac{d^5y}{dx^5}$ which occurs in $\frac{d^2R}{dx^2}$; thus (2) is a differential equation of the *fifth* order, which is a first integral of the equation (1) which is of the *sixth* order. Particular cases may be obtained by supposing R or Q or P to be zero. For example, the most useful case is that in which V involves only y and $\frac{dy}{dx}$; so that (1) becomes

$$N - \frac{dP}{dx} = 0,$$

and (2) becomes

$$V = P\frac{dy}{dx} + C.$$

334. The differential equation $K=0$ is also susceptible of one integration when V does not contain the dependent variable. For then $N=0$, and the equation becomes

$$\frac{dP}{dx} - \frac{d^2Q}{dx^2} + \frac{d^3R}{dx^3} - \ldots = 0,$$

and therefore

$$P - \frac{dQ}{dx} + \frac{d^2R}{dx^2} - \ldots = C.$$

335. We know that $\int_{x_0}^{x_1} V\,dx = \int V\frac{dx}{dy}\,dy$, supposing the limits of the integration with respect to y taken to correspond to those of the integration with respect to x. And the differential coefficients of y with respect to x may be expressed in terms of the differential coefficients of x with respect to y. Thus in $\int V\frac{dx}{dy}\,dy$ we may regard y as the independent variable, and x as the dependent variable, and proceed to find the maximum or minimum value of the integral in this new

form. We may feel *a priori* certain, as the problem is really not changed by this change of the independent variable, that we shall obtain the same result as if we had left the original independent variable.

Hence the cases considered in Arts. 333 and 334 may be seen to coincide.

336. Again, let us suppose that V involves only p and q. Then the differential equation $K=0$ reduces to

$$-\frac{dP}{dx}+\frac{d^2Q}{dx^2}=0;$$

therefore, by integration,

$$P=\frac{dQ}{dx}+C_1.$$

Also

$$\frac{dV}{dx}=P\frac{dp}{dx}+Q\frac{dq}{dx}$$

$$=C_1\frac{dp}{dx}+\frac{dQ}{dx}\frac{dp}{dx}+Q\frac{dq}{dx};$$

therefore, by integration,

$$V=Qq+C_1p+C_2.$$

Here C_1 and C_2 are arbitrary constants. In this case the differential equation $K=0$ is of the *fourth* order, and the result we have obtained is a differential equation of the *second* order; so that we have effected two steps in the integration of the differential equation $K=0$.

337. We shall now proceed to consider some examples; as we have already intimated we confine ourselves entirely to the *first part* of the process for finding maxima and minima values; see Art. 325.

338. To find the shortest line between two points.

This example is introduced merely for the purpose of illustrating the formulæ, as it is obvious that the result must be the straight line joining the two points.

Here $V = \sqrt{(1+p^2)}$ and $U = \int_{x_0}^{x_1} \sqrt{(1+p^2)}\, dx.$

Thus V involves only p, and the equation $K=0$ reduces to $\frac{dP}{dx}=0$; thus P must be a constant, that is, $\frac{p}{\sqrt{(1+p^2)}}$ must be a constant. This shews that p must be a constant, and therefore the required curve must be a straight line.

In this case $H_1 - H_0 = \frac{\delta y_1 p_1}{\sqrt{(1+p_1^2)}} - \frac{\delta y_0 p_0}{\sqrt{(1+p_0^2)}}$.

If now the two points are *fixed* points, we have $\delta y_1 = 0$ and $\delta y_0 = 0$; thus $H_1 - H_0$ vanishes. Then the value of p must be found from the condition that the straight line must pass through the two fixed points.

Suppose however that the *ordinates* of the two points are not fixed; the *abscissæ* are fixed because x_1 and x_0 are taken to be invariable. In this case δy_1 and δy_0 are arbitrary; and therefore $H_1 - H_0$ will not necessarily vanish unless the coefficients of δy_1 and δy_0 vanish. This requires that p_1 and p_0 should vanish, and as p is a constant by supposition this constant must be zero. Thus our formulæ are consistent with the obvious fact, that when two lines are parallel the shortest distance between them is obtained by drawing a line perpendicular to them both.

339. To find the curve of quickest descent from one given point to another.

The following is a fuller statement of the meaning of this problem. Suppose an indefinitely thin smooth tube connecting the two points, and a heavy particle to slide down this tube; we require to know the form of the tube in order that the time of descent may be a minimum. The problem is known by the name of the *brachistochrone;* it was first proposed by John Bernoulli in 1696, and gave rise to the Calculus of Variations.

We shall assume that the required curve lies in the vertical plane which contains the two given points. Let the axis of y be measured vertically downwards, and take the axis of

x to pass through the upper given point. The particle is supposed to start from rest, and then by the principles of mechanics the velocity at the depth y is $\sqrt{(2gy)}$. Thus the time of descent is $\int_{x_0}^{x_1} \frac{\sqrt{(1+p^2)}}{\sqrt{(2gy)}}\, dx$. We may then take

$$V = \frac{\sqrt{(1+p^2)}}{\sqrt{y}}.$$

Here V involves only y and p; so that, by Art. 333, for a minimum we must have

$$V = Pp + C,$$

that is,
$$\frac{\sqrt{(1+p^2)}}{\sqrt{y}} = \frac{p^2}{\sqrt{\{y(1+p^2)\}}} + C;$$

therefore
$$\frac{1}{\sqrt{\{y(1+p^2)\}}} = C.$$

Hence $y(1+p^2) = $ a constant $= 2a$ suppose;

therefore
$$p^2 = \frac{2a-y}{y};$$

therefore
$$\frac{dx}{dy} = \left(\frac{y}{2a-y}\right)^{\frac{1}{2}} = \frac{y}{\sqrt{(2ay-y^2)}};$$

therefore $x = a \operatorname{vers}^{-1} \frac{y}{a} - \sqrt{(2ay - y^2)} + b$, where b is another constant.

This shews that the required curve is a cycloid with its base horizontal, its vertex downwards, and a cusp at the upper point. We may suppose the origin at the upper point so that $x_0 = 0$, and then $b = 0$.

$$\text{Here } H_1 - H_0 = \left[\frac{p\delta y}{\sqrt{\{y(1+p^2)\}}}\right]_1 - \left[\frac{p\delta y}{\sqrt{\{y(1+p^2)\}}}\right]_0$$

$$= \frac{1}{\sqrt{(2a)}} \{(p\delta y)_1 - (p\delta y)_0\}.$$

As we suppose both the extreme points fixed δy_1 and δy_0 vanish, and therefore $H_1 - H_0$ vanishes.

The constant a must be determined by the condition that the cycloid shall pass through the lower given point.

Suppose however that only the abscissa of the lower point is given, and not the ordinate. Then, as before, H_0 vanishes, and $H_1 = \frac{(p\delta y)_1}{\sqrt{(2a)}}$. Now δy_1 is arbitrary, so that in order that H_1 may vanish, we must have $p_1 = 0$; thus the tangent to the cycloid at the lower limiting point must be horizontal. This condition must be used in this case to determine the constant a.

340. We may modify the preceding problem by supposing that the particle does not start from rest, but starts with an assigned velocity. In this case we will suppose that the axis of x is not drawn through the upper point, but is so taken that the velocity at starting is that which would be gained in falling from the axis of x to the upper fixed point. The solution remains as before; the cusp of the cycloid is however no longer at the upper fixed point, but in the axis of x.

341. To find the curve connecting two fixed points such that the area between the curve, its evolute, and the radii of curvature at its extremities may be a minimum.

By Art. 159 the expression which is to be made a minimum is

$$\int_{x_0}^{x_1} \frac{(1+p^2)^2}{q}\, dx.$$

Here V involves only p and q; and therefore, by Art. 336, for a minimum we must have

$$V = Qq + C_1 p + C_2,$$

that is,
$$\frac{(1+p^2)^2}{q} = -\frac{(1+p^2)^2}{q^2}q + C_1 p + C_2;$$

therefore
$$\frac{(C_1 p + C_2)\,q}{(1+p^2)^2} = 2.$$

By integration

$$C_2 \tan^{-1} p + \frac{C_2 p - C_1}{1 + p^2} = 4x + C_3 \text{ ....(1)}.$$

Also $$\frac{(C_1 p^2 + C_2 p)\, q}{(1 + p^2)^2} = 2p;$$

therefore by integration,

$$C_1 \tan^{-1} p - \frac{C_1 p + C_2}{1 + p^2} = 4y + \text{constant};$$

add C_2 to both sides of this equation, and we have

$$C_1 \tan^{-1} p + \frac{p\,(C_2 p - C_1)}{1 + p^2} = 4y + C_4 \text{ ....(2)}.$$

Eliminate $\tan^{-1} p$ from (1) and (2); thus

$$\frac{(C_2 p - C_1)^2}{1 + p^2} = 4C_2 y - 4C_1 x + C_2 C_4 - C_1 C_3,$$

therefore $$\sqrt{(1 + p^2)} = \frac{C_2 p - C_1}{2\sqrt{(C_2 y - C_1 x + B)}},$$

where B is such that $4B = C_2 C_4 - C_1 C_3$.

Let s denote the length of the arc of the curve measured from a fixed point; then, by integrating the last equation, we have

$$s + C = \sqrt{(C_2 y - C_1 x + B)}.$$

This shews that the required curve is a cycloid; see Art. 72.

We must now examine the expression $H_1 - H_0$; we have

$$H_1 = \delta y_1 \left(P - \frac{dQ}{dx}\right)_1 + \delta p_1 Q_1,$$

$$H_0 = \delta y_0 \left(P - \frac{dQ}{dx}\right)_0 + \delta p_0 Q_0.$$

As the extreme points are supposed fixed, δy_1 and δy_0 vanish; thus

$$H_1 = \delta p_1 Q_1, \quad H_0 = \delta p_0 Q_0.$$

Suppose we impose the condition that the tangents to the required curve are to have fixed directions at the extreme points; then δp_1 and δp_0 vanish, and $H_1 - H_0$ vanishes. In this case the cycloid must be determined from the conditions that it is to pass through two given points, and its tangents are to have fixed directions at these points.

If, however, no condition is imposed on the values of p at the limits, we must have $Q_1 = 0$ and $Q_0 = 0$, in order that $H_1 - H_0$ may vanish. Now $Q = -\frac{(1+p^2)^2}{q^2}$; and the radius of curvature $= \frac{(1+p^2)^{\frac{3}{2}}}{q}$. Thus the radius of curvature must vanish at the extreme points, that is, the cycloid must have cusps at those points.

342. To find the form of a solid of revolution, that the resistance on moving through a fluid in the direction of its axis may be a minimum, adopting the usual theory of resistance.

Take the axis of x as the axis of revolution. Then adopting the theory of resistance which is explained in works on Hydrodynamics, the expression which is to be a minimum is

$$\int_{x_0}^{x_1} \frac{yp^3}{1+p^2}\, dx.$$

Here V involves only y and p, and therefore by Art. 333, for a minimum we must have

$$V = Pp + C,$$

that is,
$$\frac{yp^3}{1+p^2} = y\,\frac{3p^3+p^5}{(1+p^2)^2} + C;$$

therefore
$$\frac{2yp^3}{(1+p^2)^2} + C = 0.$$

This is a differential equation for determining the required curve.

Integrals with limits subject to variation.

343. We have now sufficiently explained and illustrated the method of finding the maximum or minimum value of an integral expression involving one independent variable, when the limits of the integration are supposed invariable. We shall proceed to some extensions of the problem; and we begin by considering the modification which arises from supposing the limits of the integration variable.

Suppose, for example, that we have two given curves in one vertical plane, and that we wish to find the curve of quickest descent from one of these curves to the other, the particle starting with the velocity obtained in falling from a given horizontal line. Here we have to find the point at which the particle is to leave the upper curve, and the point of the lower curve towards which it is to proceed, as well as the path which it is to describe. We have therefore to effect more than in the examples hitherto considered, and we shall now explain how we may proceed.

We know, from what has been already given, that the curve must be a cycloid with its base horizontal and a cusp on the given horizontal line. For suppose any other curve drawn from any point in the upper curve to any point in the lower; this curve cannot be that of minimum time, for we know that, without changing the extreme points, we can find a curve of less time of descent than this curve, namely a cycloid with its base horizontal, and a cusp on the given horizontal line. Since then we know that the required curve must be such a cycloid, the part of the problem which depends on the Calculus of Variations may be considered solved; and we may investigate, by the ordinary rules for maxima and minima, the position of the particular cycloid for which the time is a minimum. In fact, taking any arbitrary initial and final points, we may find the equation to the cycloid passing through these points; then the time of descent will become a known function of the co-ordinates of the initial and final points, and we may determine for what values of these co-ordinates the time is a minimum.

344. We have shewn in the preceding article that it is not absolutely necessary to make any modification in our formulæ in order to include the case in which the limits of the integration are supposed to be susceptible of change; for the process already given, combined with the ordinary rules of the Differential Calculus, would enable us to solve any example. It is however convenient to bring together all that is wanted for solving such examples, and accordingly we shall now supply the requisite modification of our original formulæ. As before, let

$$U = \int_{x_0}^{x_1} V dx.$$

Suppose that in addition to the change of y into $y + \delta y$ the limits x_1 and x_0 are changed into $x_1 + dx_1$ and $x_0 + dx_0$ respectively. In consequence of this change of limits U receives the increment

$$\int_{x_1}^{x_1+dx_1} V dx - \int_{x_0}^{x_0+dx_0} V dx,$$

that is, neglecting squares and higher powers of dx_1 and dx_0, U receives the increment

$$V_1 dx_1 - V_0 dx_0.$$

If we annex this to the expression already given for δU, we shall obtain the complete change in U consequent upon the variation of y, and the change of the limits.

345. If no condition is imposed upon the limiting values of the co-ordinates, the additional terms just obtained,

$$V_1 dx_1 - V_0 dx_0,$$

can only be made to vanish necessarily by supposing $V_1 = 0$ and $V_0 = 0$. We thus introduce two new equations in addition to those which are obtained from $H_1 - H_0 = 0$; and at the same time we have two new quantities to determine, namely, x_0 and x_1. However, a more common case is that in which the limiting values have to satisfy given equations. Such a case we have already indicated in Art. 343, where a curve is required, the extreme points of which are to lie on given curves.

We will consider that limit of the integration for which the quantities are distinguished by the subscript 1. Let

$$Y = y + \delta y,$$

then if there had been no change of the limit, the extreme values of the variables would have been x_1 and y_1 before variation, and x_1 and Y_1 after variation. If however x_1 is changed into $x_1 + dx_1$, we have Y_1 changed into

$$\left\{Y + \frac{dY}{dx} dx_1 + \frac{1}{2}\frac{d^2 Y}{dx^2}(dx_1)^2 + \dots\right\}_1,$$

that is, neglecting squares and higher powers of dx_1 we have Y_1 changed into $Y_1 + \left(\frac{dY}{dx}\right)_1 dx_1$, that is, neglecting the product $\delta p_1 \, dx_1$, into $y_1 + \delta y_1 + \left(\frac{dy}{dx}\right)_1 dx_1$. Supposing then that the given relation which is to be satisfied by the extreme values is

$$Y = \psi(X),$$

we must have $$y_1 = \psi(x_1),$$

and also

$$y_1 + \delta y_1 + \left(\frac{dy}{dx}\right)_1 dx_1 = \psi(x_1 + dx_1) = \psi(x_1) + \psi'(x_1)\, dx_1$$

to the first order. Thus

$$\delta y_1 = \left\{\psi'(x) - \frac{dy}{dx}\right\}_1 dx_1.$$

This gives a relation between δy_1 and dx_1, so that we can eliminate one of them from the complete value of δU.

Similarly, the relation can be found between δy_0 and dx_0.

In geometrical problems $\left(\frac{dy}{dx}\right)_1$ is the tangent of the inclination to the axis of x of the line which touches the *required* curve at the limiting point; and $\psi'(x_1)$ is the tangent of the inclination to the axis of x of the line which touches the *given* curve at that point.

A particular case may be noticed which is sometimes useful. Suppose the *complete* change of y_1 is to be zero; this gives $\delta y_1 + \left(\frac{dy}{dx}\right)_1 dx_1 = 0$; similarly if the *complete* change of y_0 is to be zero, $\delta y_0 + \left(\frac{dy}{dx}\right)_0 dx_0 = 0$.

346. Let us now consider the case of the brachistochrone problem which has been enunciated in Art. 343.

Let the notation be as in Art. 339. Then

$$\delta U = V_1 dx_1 - V_0 dx_0 + \left[\frac{p\,\delta y}{\sqrt{\{y(1+p^2)\}}}\right]_1 - \left[\frac{p\,\delta y}{\sqrt{\{y(1+p^2)\}}}\right]_0$$

$$+ \int_{x_0}^{x_1} \left(N - \frac{dP}{dx}\right) \delta y\, dx.$$

As before from the equation $N - \frac{dP}{dx} = 0$ we deduce

$$\sqrt{\{y(1+p^2)\}} = \sqrt{(2a)};$$

thus $$\delta U = V_1 dx_1 - V_0 dx_0 + \frac{1}{\sqrt{(2a)}} \{(p\,\delta y)_1 - (p\,\delta y)_0\}.$$

Let us suppose that the equation to the fixed curve from which the particle is to start is $Y = \chi(X)$, and that the equation to the fixed curve at which the particle is to arrive is $Y = \psi(X)$. Then by the preceding article we have

$$\delta y_1 = \{\psi'(x) - p\}_1 dx_1, \quad \delta y_0 = \{\chi'(x) - p\}_0 dx_0.$$

Thus the value of δU can be put in the form

$$\delta U = \lambda_1 dx_1 - \lambda_0 dx_0;$$

where $$\lambda_1 = V_1 + \frac{p_1}{\sqrt{(2a)}} \{\psi'(x_1) - p_1\}$$

$$= \frac{\sqrt{(1+p_1^2)}}{\sqrt{y_1}} + \frac{p_1}{\sqrt{(2a)}} \{\psi'(x_1) - p_1\}$$

$$= \frac{1}{\sqrt{(2a)}} \{1 + p_1 \psi'(x_1)\},$$

and similarly

$$\lambda_0 = \frac{1}{\sqrt{(2a)}}\{1 + p_0 \chi'(x_0)\}.$$

Since dx_1 and dx_0 are arbitrary, δU will not necessarily vanish unless $\lambda_1 = 0$ and $\lambda_0 = 0$. Thus

$$1 + p_1 \psi'(x_1) = 0 \text{ and } 1 + p_0 \chi'(x_0) = 0;$$

and this shews that the cycloid must cut each of the two fixed curves at right angles.

347. We have hitherto tacitly assumed that the function V does not involve the limiting values of the variables or of the differential coefficients. Suppose now however that V does involve $x_0, x_1, y_0, y_1, p_0, p_1, \ldots$.

(1) Suppose that x_0 and x_1 are not susceptible of any change. When y is changed into $y + \delta y$, besides the variation we have already investigated, V will receive an additional variation arising from the change in $y_0, y_1, \ldots$ which occur explicitly in V. These additional terms in δV are

$$\frac{dV}{dy_0}\delta y_0 + \frac{dV}{dy_1}\delta y_1 + \frac{dV}{dp_0}\delta p_0 + \frac{dV}{dp_1}\delta p_1 + \ldots;$$

and consequently the following additional terms occur in δU,

$$\int_{x_0}^{x_1}\left\{\frac{dV}{dy_0}\delta y_0 + \frac{dV}{dy_1}\delta y_1 + \frac{dV}{dp_0}\delta p_0 + \frac{dV}{dp_1}\delta p_1 + \ldots\right\} dx.$$

Now $\delta y_0, \delta y_1, \delta p_0, \delta p_1, \ldots$ are not functions of the *variable* x, but only of the limiting values of x; we may therefore bring these quantities outside the integral sign and write the additional terms thus,

$$\delta y_0 \int_{x_0}^{x_1} \frac{dV}{dy_0} dx + \delta y_1 \int_{x_0}^{x_1} \frac{dV}{dy_1} dx + \delta p_0 \int_{x_0}^{x_1} \frac{dV}{dp_0} dx + \ldots$$

Now the occurrence of these additional terms will not affect the reasoning by which it is shewn in Art. 330 that we must have $K = 0$ in order that U may be a maximum or minimum. These additional terms must be annexed to the expression $H_1 - H_0$, and the whole then made to vanish. Since the relation between x and y is supposed to be found from the equation $K=0$, the expressions under the integral signs in these additional terms become definite functions of x, so that the integrations which are indicated can be effected, at least theoretically.

(2) Suppose that x_0 and x_1 are also changed, and let them become $x_0 + dx_0$ and $x_1 + dx_1$ respectively. Then V receives the additional increment

$$\left[\frac{dV}{dx_0}\right] dx_0 + \left[\frac{dV}{dx_1}\right] dx_1,$$

where $\left[\frac{dV}{dx_0}\right]$ and $\left[\frac{dV}{dx_1}\right]$ indicate *complete* differential coefficients; that is to say, we are to remember that x_0 occurs implicitly in $y_0, p_0, \ldots,$ and similarly for x_1.

Thus besides the additional terms we have already given δU receives the increment

$$dx_0 \int_{x_0}^{x_1} \left[\frac{dV}{dx_0}\right] dx + dx_1 \int_{x_0}^{x_1} \left[\frac{dV}{dx_1}\right] dx,$$

and this expression must be annexed to the aggregate formed of $H_1 - H_0$ and the additional terms already given.

348. For an example we will take another modification of the brachistochrone problem. Suppose two given curves in the same vertical plane, and let it be required to find the curve of quickest descent from one of these to the other, the motion commencing at the first curve.

Let the axis of y be measured vertically downwards; let y_0 be the ordinate of the starting point, then when the ordinate is y the velocity is $\sqrt{\{2g(y-y_0)\}}$.

Thus we may take

$$U = \int_{x_0}^{x_1} \frac{\sqrt{(1+p^2)}}{\sqrt{(y-y_0)}}\, dx.$$

We have then to change y into $y - y_0$ in the solution of Art. 346, and to add to the expression there given for δU the terms found in Article 347.

Here $V = \dfrac{\sqrt{(1+p^2)}}{\sqrt{(y-y_0)}}$; so that y_0 is the only limiting value which occurs in V. We are therefore to add to the former value of δU

$$\delta y_0 \int_{x_0}^{x_1} \frac{dV}{dy_0} dx + dx_0 \int_{x_0}^{x_1} \left[\frac{dV}{dx_0}\right] dx;$$

and
$$\left[\frac{dV}{dx_0}\right] = \frac{dV}{dy_0}\left(\frac{dy}{dx}\right)_0.$$

Hence by Art. 346, after putting $K = 0$, we have

$$\delta U = \lambda_1 dx_1 - \lambda_0 dx_0 + \left\{\delta y_0 + \left(\frac{dy}{dx}\right)_0 dx_0\right\} \int_{x_0}^{x_1} \frac{dV}{dy_0} dx,$$

where λ_1 and λ_0 have the values assigned in Art. 346.

Now in the present case

$$\frac{dV}{dy_0} = -\frac{dV}{dy} = -N = -\frac{dP}{dx},$$

therefore
$$\int_{x_0}^{x_1} \frac{dV}{dy_0} dx = P_0 - P_1 = \frac{p_0 - p_1}{\sqrt{(2a)}};$$

and
$$\delta y_0 + \left(\frac{dy}{dx}\right)_0 dx_0 = \chi'(x_0)\, dx_0, \text{ as in Art. 346.}$$

Thus
$$\delta U = \lambda_1 dx_1 - \lambda_0 dx_0 + \frac{\chi'(x_0)}{\sqrt{(2a)}}(p_0 - p_1)\, dx_0$$
$$= \frac{1}{\sqrt{(2a)}}\{1 + p_1\psi'(x_1)\}\, dx_1$$
$$- \frac{1}{\sqrt{(2a)}}\{1 + p_1\chi'(x_0)\}\, dx_0.$$

Then by equating to zero the coefficients of dx_1 and dx_0 we have

$$1 + p_1\psi'(x_1) = 0 \text{ and } 1 + p_1\chi'(x_0) = 0,$$

so that
$$\chi'(x_0) = \psi'(x_1).$$

Thus the cycloid cuts the lower fixed curve at right angles, and the tangent to the upper fixed curve at the initial point is parallel to the tangent to the lower fixed curve at the final point.

Integrals with two dependent variables.

349. We have hitherto supposed that V is a function with only one dependent variable; let us now suppose that V is a function of two dependent variables.

Let V be a function of x, y, z, and the differential coefficients of y and z with respect to x; let

$$U = \int_{x_0}^{x_1} V\,dx,$$

and let us investigate the variation in the value of U when y and z receive variations.

By proceeding as in Art. 328 we shall obtain the following result,

$$\delta U = H_1 - H_0 + J_1 - J_0 + \int_{x_0}^{x_1} (K\delta y + L\delta z)\,dx,$$

where the symbols have the following meanings;
δy, as before, denotes an arbitrary variation given to y, that is, δy is an indefinitely small arbitrary function of x;
K, as before, denotes

$$\frac{dV}{dy} - \frac{d}{dx}\frac{dV}{dy'} + \frac{d^2}{dx^2}\frac{dV}{dy''} - \ldots,$$

where $\frac{dV}{dy}$, $\frac{dV}{dy'}$, $\frac{dV}{dy''}$, ... are partial differential coefficients, and $\frac{d}{dx}\frac{dV}{dy'}$, $\frac{d^2}{dx^2}\frac{dV}{dy''}$, ... are *complete* differential coefficients relative to x;

δz is an arbitrary variation given to z, that is, δz is an indefinitely small arbitrary function of x;
L is relatively to z the same as K relatively to y, that is,

$$L = \frac{dV}{dz} - \frac{d}{dx}\frac{dV}{dz'} + \frac{d^2}{dx^2}\frac{dV}{dz''} - \ldots;$$

$H_1 - H_0$ has the meaning already given, and $J_1 - J_0$ is relatively to z the same as $H_1 - H_0$ relatively to y.

350. We now proceed to find a maximum or minimum value of U on the suppositions of the preceding article.

(1) If y and z are independent, in order that δU may certainly vanish we must have

$$K = 0 \text{ and } L = 0;$$

and also
$$H_1 - H_0 + J_1 - J_0 = 0.$$

The values of y and z in terms of x must be found by solving the differential equations $K = 0$, $L = 0$; and the arbitrary constants which occur in these solutions must be determined by equating to zero the coefficients of the arbitrary quantities δy_0, δy_1, $\left(\delta \frac{dy}{dx}\right)_0$, ... δz_0, δz_1, $\left(\delta \frac{dz}{dx}\right)_0$, ... which occur in $H_1 - H_0 + J_1 - J_0$.

(2) Suppose however that y and z are not independent, but that they are connected by the relation $\phi(x, y, z) = 0$, which is always to hold. Since this relation is supposed to hold always, we have also

$$\phi(x, y + \delta y, z + \delta z) = 0;$$

and therefore ultimately

$$\frac{d\phi}{dy} \delta y + \frac{d\phi}{dz} \delta z = 0.$$

Thus the integral $\int_{x_0}^{x_1} (K \delta y + L \delta z)\, dx$ becomes

$$\int_{x_0}^{x_1} \left\{ K - \frac{L \dfrac{d\phi}{dy}}{\dfrac{d\phi}{dz}} \right\} \delta y\, dx,$$

and in order that this may vanish we have the single condition

$$\frac{K}{\dfrac{d\phi}{dy}} = \frac{L}{\dfrac{d\phi}{dz}};$$

and from this differential equation combined with $\phi(x, y, z)=0$, we must find y and z.

As before, we must also have

$$H_1 - H_0 + J_1 - J_0 = 0.$$

351. For an example we take the following problem; to determine a line of minimum length on a given curved surface, between two given points.

Here we have

$$U=\int_{x_0}^{x_1}\sqrt{\left\{1+\left(\frac{dy}{dx}\right)^2+\left(\frac{dz}{dx}\right)^2\right\}}\,dx=\int_{x_0}^{x_1}\sqrt{(1+y'^2+z'^2)}\,dx;$$

thus $$K=-\frac{d}{dx}\frac{y'}{\sqrt{(1+y'^2+z'^2)}},\quad L=-\frac{d}{dx}\frac{z'}{\sqrt{(1+y'^2+z'^2)}};$$

let $\phi(x, y, z)=0$ be the equation to the surface on which the line lies. Then by the preceding article we have, as the condition for a minimum,

$$\frac{\dfrac{d}{dx}\dfrac{y'}{\sqrt{(1+y'^2+z'^2)}}}{\dfrac{d\phi}{dy}}=\frac{\dfrac{d}{dx}\dfrac{z'}{\sqrt{(1+y'^2+z'^2)}}}{\dfrac{d\phi}{dz}}.$$

Let s represent the length of the arc of the curve; then

$$\frac{y'}{\sqrt{(1+y'^2+z'^2)}}=\frac{dy}{ds},\text{ and }\frac{z'}{\sqrt{(1+y'^2+z'^2)}}=\frac{dz}{ds}.$$

Thus the above equation may be written

$$\frac{\dfrac{d^2y}{ds^2}}{\dfrac{d\phi}{dy}}=\frac{\dfrac{d^2z}{ds^2}}{\dfrac{d\phi}{dz}}\quad\text{.........................}\quad(1).$$

From this we may conjecture by symmetry that each of these fractions is equal to

$$\frac{\dfrac{d^2x}{ds^2}}{\dfrac{d\phi}{dx}},$$

and this we can prove; for from (1) each of the fractions by a known theorem of algebra is equal to

$$\frac{\frac{dy}{ds}\frac{d^2y}{ds^2}+\frac{dz}{ds}\frac{d^2z}{ds^2}}{\frac{dy}{ds}\frac{d\phi}{dy}+\frac{dz}{ds}\frac{d\phi}{dz}};$$

and since the equation $\phi(x, y, z)=0$ holds for every point of the curve, we have

$$\frac{d\phi}{dx}\frac{dx}{ds}+\frac{d\phi}{dy}\frac{dy}{ds}+\frac{d\phi}{dz}\frac{dz}{ds}=0;$$

also by a known theorem

$$\frac{dx}{ds}\frac{d^2x}{ds^2}+\frac{dy}{ds}\frac{d^2y}{ds^2}+\frac{dz}{ds}\frac{d^2z}{ds^2}=0.$$

Hence a line of minimum length is determined by the symmetrical equations

$$\frac{\frac{d^2x}{ds^2}}{\frac{d\phi}{dx}}=\frac{\frac{d^2y}{ds^2}}{\frac{d\phi}{dy}}=\frac{\frac{d^2z}{ds^2}}{\frac{d\phi}{dz}} \qquad (2).$$

It is proved in works on Geometry of Three Dimensions that the equations (2) shew that the osculating plane at any point of the curve contains the normal to the surface at that point.

Relative Maxima and Minima.

352. A class of problems still remains to be considered, called problems of *relative* maxima and minima values. Suppose we require that a certain integral U shall have a maximum or minimum value while another integral W, involving the same variables, has a constant value; for example, we may require a curve which shall include a minimum area under a given perimeter. Here we do not require that δU shall always vanish, but only that it shall vanish for such relations among the variables as give a definite constant value

to W; that is in fact, we require that δU shall vanish for all such relations among the variables as make δW vanish.

The problem is solved by finding a maximum or minimum value of $U + aW$, where a denotes a constant; for in this solution we ensure that $\delta U + a\delta W$ necessarily vanishes, and therefore δU must vanish whenever δW does. The constant a occurs in the solution, and its value must be determined by making the integral W have the constant value which is supposed given.

If we require that W shall be a maximum or minimum while U remains constant, we shall in the same way proceed to find the maximum or minimum of $W + bU$, where b is a constant; and if we suppose $b = \frac{1}{a}$, we obtain the expression $\frac{1}{a}(U + aW)$. Thus the same solution will be obtained for this problem as for that in which U is to be a maximum or minimum while W is constant.

We now proceed to some examples.

353. It is required to find a curve of given length joining two fixed points, so that the area bounded by the curve, the axis of x, and ordinates at the fixed points may be a maximum.

Here $U = \int_{x_0}^{x_1} y\,dx, \quad W = \int_{x_0}^{x_1} \sqrt{(1 + p^2)}\,dx$;

let $V = y + a\sqrt{(1 + p^2)}$, then we have to investigate a maximum or minimum value of $\int_{x_0}^{x_1} V dx$. Under the integral sign we have only y and p; hence for a maximum or minimum, by Art. 333, we must have

$$V = Pp + C_1,$$

that is,
$$y + a\sqrt{(1 + p^2)} = \frac{ap^2}{\sqrt{(1 + p^2)}} + C_1,$$

that is,
$$y + \frac{a}{\sqrt{(1 + p^2)}} = C_1.$$

Thus $$1+p^2=\frac{a^2}{(C_1-y)^2},$$

therefore $$\left(\frac{dx}{dy}\right)^2=\frac{1}{p^2}=\frac{(C_1-y)^2}{a^2-(C_1-y)^2};$$

therefore $$x+C_2=\sqrt{\{a^2-(C_1-y)^2\}}.$$

This shews that the required curve is a circular arc.

Since the extreme points are supposed fixed, the part of δV which depends on the limits vanishes.

The constants C_1, C_2, a must be determined by making the circular arc pass through the given fixed points and have the given length between them.

354. Given the length of a curve, find its form so that the depth of the centre of gravity may be a maximum.

Take the axis of x horizontal, and the axis of y vertically downwards. Let b denote the length of the curve; then the depth of the centre of gravity is $\frac{1}{b}\int_{x_0}^{x_1} y\sqrt{(1+p^2)}\,dx$, and the length is $\int_{x_0}^{x_1}\sqrt{(1+p^2)}\,dx$.

Let $$V=\frac{1}{b}y\sqrt{(1+p^2)}+a\sqrt{(1+p^2)},$$

then we require a maximum or minimum value of $\int_{x_0}^{x_1} V\,dx$.

Here by Art. 333 we must have

$$V=Pp+C_1, \text{ that is,}$$

$$\frac{y}{b}\sqrt{(1+p^2)}+a\sqrt{(1+p^2)}=\frac{p^2y}{b\sqrt{(1+p^2)}}+\frac{ap^2}{\sqrt{(1+p^2)}}+C_1,$$

therefore $$\frac{y+ab}{\sqrt{(1+p^2)}}=bC_1;$$

therefore $$1+p^2=\frac{(y+ab)^2}{b^2C_1^2},$$

and therefore
$$\frac{dx}{dy} = \frac{bC_1}{\sqrt{\{(y+ab)^2 - b^2C_1^2\}}};$$

hence $\quad x = A \log [y + B + \sqrt{\{(y+B)^2 - A^2\}}] + C_2,$

where C_2 is a new constant, and $A = bC_1$ and $B = ab$.

This equation shews that the required curve is a catenary. If the ends of the required curve are supposed fixed, the terms depending on the limits vanish, and the constants A, B, C_2 must be determined by making the catenary pass through the fixed points and have a given length between them. Suppose however that instead of being fixed the ends are only constrained to lie on fixed curves. By proceeding as in Art. 346 we obtain the following limiting terms;

$$V_1 dx_1 - V_0 dx_0 + P_1 \delta y_1 - P_0 \delta y_0.$$

Consider the terms with the suffix 1; we have $V_1 dx_1 + P_1 \delta y_1$,

that is, $\left(\frac{y_1}{b} + a\right) \sqrt{(1+p_1^2)}\, dx_1 + \left(\frac{y_1}{b} + a\right) \frac{p_1 \delta y_1}{\sqrt{(1+p_1^2)}}.$

Now supposing $y = \psi(x)$ the equation to the fixed curve, we have $\delta y_1 = \{\psi'(x_1) - p_1\} dx_1$, so that the term reduces to

$$\frac{y_1 + ab}{b\sqrt{(1+p_1^2)}} \{1 + p_1 \psi'(x_1)\} dx_1.$$

To make this vanish we must have $1 + p_1\psi'(x_1) = 0$, for $y_1 + ab$ cannot vanish, as then x_1 would be impossible. A similar result holds at the other limit; and thus it appears that the catenary must cut the fixed curves at right angles.

355. Given the surface of a solid of revolution, to find its nature that the solid content may be a maximum.

Take the axis of x as the axis of revolution. Then the surface is $2\pi \int_{x_0}^{x_1} y\sqrt{(1+p^2)}\, dx$, and the volume is $\pi \int_{x_0}^{x_1} y^2\, dx$.

Let $V = y^2 + ay\sqrt{(1+p^2)}$; then we have to find a maximum or minimum value of $\int_{x_0}^{x_1} V\, dx$. Here by Art. 333 we must have

$$V = Pp + C,$$

that is, $$y^2 + ay\sqrt{(1+p^2)} = \frac{ayp^2}{\sqrt{(1+p^2)}} + C,$$

therefore $$y^2 + \frac{ay}{\sqrt{(1+p^2)}} = C.$$

This is a differential equation to the curve which would by revolution generate the required surface. Supposing that the ends of the generating curve are required to pass through fixed points, the terms at the limits vanish.

If either of the fixed points is on the axis of revolution, the value $y = 0$ is to satisfy the equation to the curve; thus $C = 0$. Then the general equation reduces to

$$y^2 + \frac{ay}{\sqrt{(1+p^2)}} = 0, \quad \text{therefore } y + \frac{a}{\sqrt{(1+p^2)}} = 0;$$

this gives a circular arc as the generating curve.

Some further discussion of this problem will be found in the *Philosophical Magazine* for July and for August 1861.

356. Given the mass of a solid of revolution of uniform density, required its form so that its attraction upon a point in its axis may be a maximum.

Let the axis of x be taken as that of revolution, and the position of the attracted point as the origin.

Let the solid be divided into indefinitely thin slices by planes perpendicular to the axis of x. If y represent the radius of a slice, x its distance from the attracted point, κ its thickness and ρ its density, the attraction is (see *Statics*, Chapter XIII.)

$$2\pi\rho\kappa\left\{1 - \frac{x}{\sqrt{(x^2+y^2)}}\right\}.$$

Therefore the whole attraction of the solid is

$$2\pi\rho\int_{x_0}^{x_1}\left\{1 - \frac{x}{\sqrt{(x^2+y^2)}}\right\}dx;$$

and the mass of the solid is

$$\pi\rho\int_{x_0}^{x_1} y^2\,dx.$$

Thus let $V = 1 - \frac{x}{\sqrt{(x^2+y^2)}} + ay^2$; then we have to investigate the maximum or minimum value of $\int_{x_0}^{x_1} V\,dx$.

The condition $N - \frac{dP}{dx} + \ldots\ldots = 0$ reduces here to $N = 0$,

that is, $$2ay + \frac{xy}{(x^2+y^2)^{\frac{3}{2}}} = 0;$$

therefore $$2a\,(x^2+y^2)^{\frac{3}{2}} + x = 0.$$

If we suppose the limits x_1 and x_0 susceptible of change we have the limiting terms $V_1 dx_1 - V_0 dx_0$; and to make these vanish we must have $V_1 = 0$ and $V_0 = 0$; this leads to $y_1 = 0$ and $y_0 = 0$. Thus the solid must be formed by the revolution round the axis of x of the whole closed curve determined by the equation $2a\,(x^2+y^2)^{\frac{3}{2}} + x = 0$; the value of a must be found from the condition that the mass, and therefore the volume, is given.

Double Integrals.

357. We shall now consider the problem of finding the maximum or minimum value of a *double integral;* and we begin by finding the variation of a double integral.

Let z be a function of the independent variables x and y at present unknown; let V be a given function of x, y, z, $\frac{dz}{dx}$ and $\frac{dz}{dy}$; let $U = \int_{x_0}^{x_1}\int_{y_0}^{y_1} V\,dx\,dy$; the integration is supposed effected with respect to y first, and the limits y_0 and y_1 are supposed given functions of x. It is required to determine what function z must be of x and y in order that U may have a maximum or minimum value.

Let δz denote an indefinitely small arbitrary function of x and y; let δV denote the variation made in V when z receives the variation δz, and let δU denote the variation in U; then we have first to obtain an expression for δU.

Let L denote the partial differential coefficient of V with respect to z, M the partial differential coefficient of V with respect to $\frac{dz}{dx}$, and N the partial differential coefficient of V with respect to $\frac{dz}{dy}$; then we have

$$\delta V = L\delta z + M\frac{d\,\delta z}{dx} + N\frac{d\,\delta z}{dy},$$

where, as heretofore, we confine ourselves to the first power of the indefinitely small quantities. Hence

$$\delta U = \int_{x_0}^{x_1}\int_{y_0}^{y_1}\left(L\delta z + M\frac{d\delta z}{dx} + N\frac{d\,\delta z}{dy}\right) dx\, dy.$$

The value of δV may be written thus;

$$\delta V = \left(L - \frac{dM}{dx} - \frac{dN}{dy}\right)\delta z + \frac{d}{dx}(M\delta z) + \frac{d}{dy}(N\delta z),$$

and therefore

$$\delta U = \int_{x_0}^{x_1}\int_{y_0}^{y_1}\left(L - \frac{dM}{dx} - \frac{dN}{dy}\right)\delta z\, dx\, dy$$

$$+ \int_{x_0}^{x_1}\int_{y_0}^{y_1}\frac{d}{dx}(M\delta z)\, dx\, dy + \int_{x_0}^{x_1}\int_{y_0}^{y_1}\frac{d}{dy}(N\delta z)\, dx\, dy.$$

The differential coefficients with respect to x and y which are here indicated are *complete* differential coefficients.

Also $\int_{x_0}^{x_1}\int_{y_0}^{y_1}\frac{d}{dy}(N\delta z)\, dx\, dy = \int_{x_0}^{x_1}\{(N\delta z)_1 - (N\delta z)_0\}\, dx,$

where $(N\delta z)_1$ denotes the value of $N\delta z$ when y_1 is put for y, and $(N\delta z)_0$ denotes the value of $N\delta z$ when y_0 is put for y.

And by Art. 216,

$$\int_{y_0}^{y_1}\frac{d}{dx}(M\delta z)\, dy = \frac{d}{dx}\int_{y_0}^{y_1} M\delta z\, dy - (M\delta z)_1\frac{dy_1}{dx} + (M\delta z)_0\frac{dy_0}{dx},$$

where $(M\delta z)_1$ denotes the value of $M\delta z$ when y_1 is put for y, and $(M\delta z)_0$ denotes the value of $M\delta z$ when y_0 is put for y.

Therefore $$\int_{x_0}^{x_1}\int_{y_0}^{y_1}\frac{d}{dx}(M\delta z)\,dx\,dy$$

$$=\left(\int_{y_0}^{y_1} M\delta z\,dy\right)_{x=x_1}-\left(\int_{y_0}^{y_1} M\delta z\,dy\right)_{x=x_0}$$

$$-\int_{x_0}^{x_1}(M\delta z)_1\frac{dy_1}{dx}\,dx+\int_{x_0}^{x_1}(M\delta z)_0\frac{dy_0}{dx}\,dx.$$

Therefore $$\delta U=\int_{x_0}^{x_1}\int_{y_0}^{y_1}\left(L-\frac{dM}{dx}-\frac{dN}{dy}\right)\delta z\,dx\,dy$$

$$+\int_{x_0}^{x_1}\left\{(N\delta z)_1-(N\delta z)_0\right\}dx$$

$$+\left(\int_{y_0}^{y_1} M\delta z\,dy\right)_{x=x_1}-\left(\int_{y_0}^{y_1} M\delta z\,dy\right)_{x=x_0}$$

$$-\int_{x_0}^{x_1}(M\delta z)_1\frac{dy_1}{dx}\,dx+\int_{x_0}^{x_1}(M\delta z)_0\frac{dy_0}{dx}\,dx.$$

If the limits y_1 and y_0 are *constants*, the last two terms vanish.

358. In the value of δU found in the preceding article, there is one term which is a double integral involving δz under the integral signs, and various single integrals depending upon the limiting values of δz. By the method already used in Art. 330, it will follow that δU will not certainly vanish unless the coefficient of δz under the double integral sign vanishes; thus for a maximum or minimum value of U we have as a necessary condition

$$L-\frac{dM}{dx}-\frac{dN}{dy}=0.$$

This is a partial differential equation for finding z in terms of x and y; and we may say that the arbitrary functions which occur in its solution must be determined so that the remaining terms in δU may vanish. But the difficulty of integrating the partial differential equation in general prevents any practical examination of these terms at the limits.

359. As an example, let it be required to determine a surface of minimum area bounded by a given curve.

Here by Art. 170,

$$U = \int_{x_0}^{x_1}\int_{y_0}^{y_1} \sqrt{\left\{1 + \left(\frac{dz}{dx}\right)^2 + \left(\frac{dz}{dy}\right)^2\right\}}\, dx\, dy;$$

let us put as usual

$$\frac{dz}{dx} = p, \quad \frac{dz}{dy} = q, \quad \frac{d^2z}{dx^2} = r, \quad \frac{d^2z}{dx\,dy} = s, \quad \frac{d^2z}{dy^2} = t.$$

The condition for a minimum reduces to

$$\frac{dM}{dx} + \frac{dN}{dy} = 0,$$

that is, to $$\frac{d}{dx}\frac{p}{\sqrt{(1+p^2+q^2)}} + \frac{d}{dy}\frac{q}{\sqrt{(1+p^2+q^2)}} = 0,$$

that is, to

$$r(1+p^2+q^2) - (pr+qs)\,p + t(1+p^2+q^2) - (ps+qt)\,q = 0,$$

that is, to $$(1+q^2)\,r - 2pqs + (1+p^2)\,t = 0.$$

It is proved in works on Geometry of Three Dimensions that this equation indicates that the required surface is such that at every point the two principal radii of curvature are equal in magnitude and of contrary signs.

Since we suppose the boundary of the required surface to be a fixed curve δz vanishes all round this boundary; thus the terms relative to the limits in δU all vanish.

Discrimination of Maxima and Minima values.

360. We shall now make some remarks on the second part of the investigation of maxima and minima values of integrals; see Art. 325.

Consider the example of finding the shortest line between two given points. Here

$$V = \sqrt{(1+p^2)}, \qquad U = \int_{x_0}^{x_1} V\,dx.$$

Suppose y changed into $y + \delta y$, and consequently p into $p + \delta p$; put $p + \delta p$ instead of p in V and expand; thus V becomes

$$\sqrt{(1+p^2)} + \frac{p\,\delta p}{\sqrt{(1+p^2)}} + \frac{(\delta p)^2}{2\,(1+p^2)^{\frac{3}{2}}} - \dots,$$

where the terms which are not expressed are of the *third* and higher orders in δp. Thus we obtain

$$\delta U = \int_{x_0}^{x_1} \frac{p\,\delta p}{\sqrt{(1+p^2)}}\,dx + \frac{1}{2}\int_{x_0}^{x_1} \frac{(\delta p)^2}{(1+p^2)^{\frac{3}{2}}}\,dx - \dots.$$

The first of these terms is what we formerly denoted by δU, and the investigation of the minimum value of U so far as it has hitherto been carried, consists in making this term vanish. Supposing then that this term vanishes, and neglecting terms of the third and higher orders, we have

$$\delta U = \frac{1}{2}\int_{x_0}^{x_1} \frac{(\delta p)^2}{(1+p^2)^{\frac{3}{2}}}\,dx.$$

If $x_1 - x_0$ is positive, every element of this integral is positive; thus δU is *positive*, and therefore a *minimum* value of U has been obtained.

361. Again, take the case of the brachistochrone, when the extreme points are fixed. Here

$$V = \frac{\sqrt{(1+p^2)}}{\sqrt{y}}, \qquad U = \int_{x_0}^{x_1} \frac{\sqrt{(1+p^2)}}{\sqrt{y}}\,dx.$$

Change y into $y + \delta y$, and p into $p + \delta p$; and expand the new value of V. Thus V becomes

$$\frac{\sqrt{(1+p^2)}}{\sqrt{y}} - \frac{\sqrt{(1+p^2)}\,\delta y}{2y^{\frac{3}{2}}} + \frac{p\,\delta p}{y^{\frac{1}{2}}(1+p^2)^{\frac{1}{2}}}$$

$$+ \frac{3(1+p^2)^{\frac{1}{2}}(\delta y)^2}{8y^{\frac{5}{2}}} - \frac{p\,\delta y\,\delta p}{2y^{\frac{3}{2}}(1+p^2)^{\frac{1}{2}}} + \frac{(\delta p)^2}{2y^{\frac{1}{2}}(1+p^2)^{\frac{3}{2}}}$$

$$-\ldots\ldots;$$

and from this we can obtain δU.

Now by the process of Art. 339 the terms of the first order in δU are made to vanish; then, neglecting terms of the third and higher orders, we have

$$\delta U = \int_{x_0}^{x_1} \left\{ \frac{3(1+p^2)^{\frac{1}{2}}(\delta y)^2}{8y^{\frac{5}{2}}} - \frac{p\,\delta y\,\delta p}{2y^{\frac{3}{2}}(1+p^2)^{\frac{1}{2}}} + \frac{(\delta p)^2}{2y^{\frac{1}{2}}(1+p^2)^{\frac{3}{2}}} \right\} dx.$$

We have now to investigate the sign of this expression when the relation between x and y is that which is determined in Art. 339; and we shall shew by some transformations that δU is positive.

Since $$y^{\frac{1}{2}}(1+p^2)^{\frac{1}{2}} = (2a)^{\frac{1}{2}},$$

we have $$\delta U = \int_{x_0}^{x_1} \left\{ \frac{3(2a)^{\frac{1}{2}}(\delta y)^2}{8y^3} - \frac{p\,\delta y\,\delta p}{2y(2a)^{\frac{1}{2}}} + \frac{y(\delta p)^2}{4a(2a)^{\frac{1}{2}}} \right\} dx$$

$$= \frac{1}{2(2a)^{\frac{1}{2}}} \int_{x_0}^{x_1} \left\{ \frac{3a(\delta y)^2}{2y^3} - \frac{p\,\delta y\,\delta p}{y} + \frac{y(\delta p)^2}{2a} \right\} dx.$$

Now $$\int \frac{p\,\delta y\,\delta p}{y}\,dx = \frac{p(\delta y)^2}{2y} - \frac{1}{2}\int (\delta y)^2 \frac{d}{dx}\left(\frac{p}{y}\right) dx;$$

and as the extreme points are supposed fixed δy vanishes at the limits; therefore

$$\int_{x_0}^{x_1} \frac{p\,\delta y\,\delta p}{y}\,dx = -\frac{1}{2}\int_{x_0}^{x_1} (\delta y)^2 \frac{d}{dx}\left(\frac{p}{y}\right) dx.$$

Now $$\frac{d}{dx}\left(\frac{p}{y}\right) = \frac{1}{y}p\frac{dp}{dy} - \frac{p^2}{y^2} = -\frac{a}{y^3} - \frac{p^2}{y^2} = -\frac{3a-y}{y^3}.$$

Therefore $$\int_{x_0}^{x_1} \frac{p\,\delta y\,\delta p}{y}\,dx = \frac{1}{2}\int_{x_0}^{x_1} (\delta y)^2 \frac{3a-y}{y^3}\,dx;$$

and $$\delta U = \frac{1}{2(2a)^{\frac{1}{2}}} \int_{x_0}^{x_1} \left\{ \frac{(\delta y)^2}{2y^2} + \frac{y(\delta p)^2}{2a} \right\} dx.$$

Thus δU is positive, and therefore a minimum value of U has been obtained.

362. The preceding article shews that it may be possible to change the expression of the second order to which δU is reduced by our previous investigations, from a form in which the sign is uncertain to a form in which the sign is obvious. A general theory with respect to suitable transformations of such terms of the second order has been given by Jacobi; for this we refer to the works named at the end of the present chapter.

It may be observed that many of the problems discussed in the Calculus of Variations are of a kind in which we may infer with more or less certainty, from the nature of the particular problem, that there can be a minimum and no maximum, or a maximum and no minimum.

363. In the problem discussed in Art. 359 it is easy to shew that the result really gives a minimum. Here

$$V = \sqrt{(1 + p^2 + q^2)}, \quad U = \int_{x_0}^{x_1} \int_{y_0}^{y_1} \sqrt{(1 + p^2 + q^2)}\, dx\, dy.$$

Suppose z changed into $z + \delta z$, in consequence of which p becomes $p + \delta p$ and q becomes $q + \delta q$. Thus V becomes

$$(1 + p^2 + q^2)^{\frac{1}{2}} + \frac{p\delta p}{(1 + p^2 + q^2)^{\frac{1}{2}}} + \frac{q\delta q}{(1 + p^2 + q^2)^{\frac{1}{2}}}$$

$$+ \frac{(1 + q^2)(\delta p)^2}{2(1 + p^2 + q^2)^{\frac{3}{2}}} - \frac{pq\,\delta p\,\delta q}{(1 + p^2 + q^2)^{\frac{3}{2}}} + \frac{(1 + p^2)(\delta q)^2}{2(1 + p^2 + q^2)^{\frac{3}{2}}}$$

$$- \dots\dots\dots\dots$$

Then supposing the terms of the first order made to vanish, and neglecting terms of the third and higher orders, we have

$$\delta U = \frac{1}{2}\int_{x_0}^{x_1}\int_{y_0}^{y_1} \frac{(1+q^2)(\delta p)^2 - 2pq\,\delta p\,\delta q + (1+p^2)(\delta q)^2}{(1+p^2+q^2)^{\frac{3}{2}}}\,dx\,dy$$

$$= \frac{1}{2}\int_{x_0}^{x_1}\int_{y_0}^{y_1} \frac{(\delta p)^2 + (\delta q)^2 + (q\,\delta p - p\,\delta q)^2}{(1+p^2+q^2)^{\frac{3}{2}}}\,dx\,dy.$$

Thus the term under the integral signs is necessarily positive; so that a minimum value of U has been obtained.

Condition of Integrability.

364. In Art. 330 we have found that $K=0$ is a necessary condition for the existence of a maximum or minimum value of the integral there considered. It may however happen that in certain cases the relation $K=0$ is satisfied *identically;* this case we proceed to exemplify and interpret.

Suppose we are seeking a maximum or minimum value of

$$\int_{x_0}^{x_1}\left(\frac{y'}{y} - \frac{xy'^2}{y^2} + \frac{xy''}{y}\right)dx.$$

Here
$$V = \frac{y'}{y} - \frac{xy'^2}{y^2} + \frac{xy''}{y},$$

$$N = \frac{dV}{dy} = -\frac{y'}{y^2} + \frac{2xy'^2}{y^3} - \frac{xy''}{y^2},$$

$$P = \frac{dV}{dy'} = \frac{1}{y} - \frac{2xy'}{y^2},$$

$$Q = \frac{dV}{dy''} = \frac{x}{y};$$

$$N - \frac{dP}{dx} + \frac{d^2Q}{dx^2} = -\frac{y'}{y^2} + \frac{2xy'^2}{y^3} - \frac{xy''}{y^2}$$

$$-\left\{-\frac{y'}{y^2} - \frac{2y'}{y^2} - \frac{2xy''}{y^2} + \frac{4xy'^2}{y^3}\right\}$$

$$-\frac{2y'}{y^2} - \frac{xy''}{y^2} + \frac{2xy'^2}{y^3}.$$

On collecting the terms it will be found that

$$N - \frac{dP}{dx} + \frac{d^2Q}{dx^2}$$

vanishes. Thus the relation $K = 0$ is an *identity* in this example, and we cannot obtain from it any value of y.

In this example we shall find that

$$\int V dx = \frac{xy'}{y},$$

that is, the integral $\int V dx$ can be obtained without assigning the value of y in terms of x. Thus if we wish to find a maximum or minimum value of $\int_{x_0}^{x_1} V dx$, we must investigate a maximum or minimum value of $\left(\frac{xy'}{y}\right)_1 - \left(\frac{xy'}{y}\right)_0$. We are therefore not concerned with the maximum or minimum of an undetermined integral expression of the kind hitherto considered, but with the maximum or minimum of an expression free from the integral sign.

This species of maximum and minimum problem is considered in some of the comprehensive treatises on the Calculus of Variations; as it does not present much interest we will refer the student to such works.

365. We shall now prove universally that the necessary and sufficient condition in order that V may be integrable without assigning the specific value of y in terms of x, is that $K = 0$ should be identically true. An expression which is integrable without assigning the specific value of the dependent variable in terms of the independent variable is sometimes said to be integrable *per se*, and is sometimes said to be *immediately integrable*.

366. We first prove that the condition is necessary. Suppose that V involves x, y and the differential coefficients of y with respect to x up to $\frac{d^n y}{dx^n}$ inclusive.

If the function V is immediately integrable the integral $\int_{x_0}^{x_1} Vdx$ can be expressed in the form

$$\phi\left\{x_1, y_1, \left(\frac{dy}{dx}\right)_1, \left(\frac{d^2y}{dx^2}\right)_1, \ldots\ldots \left(\frac{d^{n-1}y}{dx^{n-1}}\right)_1\right\}$$

$$-\phi\left\{x_0, y_0, \left(\frac{dy}{dx}\right)_0, \left(\frac{d^2y}{dx^2}\right)_0, \ldots\ldots \left(\frac{d^{n-1}y}{dx^{n-1}}\right)_0\right\},$$

where the form of the function denoted by ϕ remains unchanged whatever may be the value of y in terms of x. Now suppose that y receives such a variation as leaves the values of y and its differential coefficients *at the limits* unaltered; then from the value of $\int_{x_0}^{x_1} Vdx$ it follows that

$$\delta\int_{x_0}^{x_1} Vdx = 0;$$

thus by Art. 329

$$\int_{x_0}^{x_1} \delta y \left\{\frac{dV}{dy} - \frac{d}{dx}\frac{dV}{dy'} + \frac{d^2}{dx^2}\frac{dV}{dy''} - \ldots\right\} dx = 0.$$

But this cannot be true whatever δy may be, unless

$$\frac{dV}{dy} - \frac{d}{dx}\frac{dV}{dy'} + \frac{d^2}{dx^2}\frac{dV}{dy''} - \ldots = 0,$$

and unless this is *identically* true it determines y as a function of x. Thus if V is immediately integrable the relation $K=0$ must be identically true.

Next we shall shew conversely that if this condition holds V *is* immediately integrable. It is usually considered sufficient to say, that if this condition holds the *variation* of $\int_{x_0}^{x_1} Vdx$ depends solely on the *limiting values* of x, y, and the differential coefficients of y; and therefore $\int_{x_0}^{x_1} Vdx$ must itself depend solely on these limiting values, that is, V must be immediately integrable. We shall however reproduce a more satisfactory demonstration which has been given of the proposition.

Suppose $V = \phi(x, y, y', y'', \ldots)$.

Let u and v denote two functions of x at present undetermined; let α denote a quantity which we shall vary independently of x. Let $\psi(\alpha)$ denote what V becomes when we put $u + \alpha v$ instead of y, and $u' + \alpha v'$ instead of y', and $u'' + \alpha v''$ instead of y'', and so on; thus

$$\psi(\alpha) = \phi(x, u + \alpha v, u' + \alpha v', u'' + \alpha v'', \ldots).$$

Differentiate both sides with respect to α, so that we have a result which we may denote thus,

$$\psi'(\alpha) = \frac{d\phi}{du} v + \frac{d\phi}{du'} v' + \frac{d\phi}{du''} v'' + \ldots.$$

Integrate both sides, from $\alpha = 0$ to $\alpha = 1$; thus

$$\psi(1) - \psi(0) = \int_0^1 \left\{\frac{d\phi}{du} v + \frac{d\phi}{du'} v' + \frac{d\phi}{du''} v'' + \ldots\right\} d\alpha;$$

that is, we have the following identically true,

$$\phi(x, u + v, u' + v', u'' + v'', \ldots)$$

$$= \phi(x, u, u', u'', \ldots)$$

$$+ \int_0^1 \left\{\frac{d\phi}{du} v + \frac{d\phi}{du'} v' + \frac{d\phi}{du''} v'' + \ldots\right\} d\alpha.$$

Integrate both sides with respect to x; thus

$$\int \phi(x, u + v, u' + v', u'' + v'', \ldots)\, dx$$

$$= \int \phi(x, u, u', u'', \ldots)\, dx$$

$$+ \int_0^1 d\alpha \left[\int \left\{\frac{d\phi}{du} v + \frac{d\phi}{du'} v' + \frac{d\phi}{du''} v'' + \ldots\right\} dx\right],$$

where in the last term the order of the independent integrations has been changed.

By integration by parts

$$\int \frac{d\phi}{du'} v' \, dx = v \frac{d\phi}{du'} - \int v \frac{d}{dx} \frac{d\phi}{du'} \, dx,$$

$$\int \frac{d\phi}{du''} v'' dx = v' \frac{d\phi}{du''} - v \frac{d}{dx} \frac{d\phi}{du''} + \int v \frac{d^2}{dx^2} \frac{d\phi}{du''} \, dx,$$

and so on.

Thus

$$\int \phi \left(x, \; u + v, \; u' + v', \; u'' + v'', \ldots\right) dx$$

$$= \int \phi \left(x, \; u, \; u', \; u'', \ldots\right) dx$$

$$+ \int_0^1 v \left(\frac{d\phi}{du'} - \frac{d}{dx} \frac{d\phi}{du''} + \ldots\right) d\alpha$$

$$+ \int_0^1 v' \left(\frac{d\phi}{du''} - \ldots\right) d\alpha$$

$$+ \quad \ldots\ldots\ldots\ldots\ldots\ldots\ldots\ldots$$

$$+ \int_0^1 d\alpha \left[\int v \left\{\frac{d\phi}{du} - \frac{d}{dx} \frac{d\phi}{du'} + \frac{d^2}{dx^2} \frac{d\phi}{du''} - \ldots\right\} dx\right].$$

Now by supposition the relation $K = 0$ is satisfied identically whatever may be the value of y; so it is satisfied if $u + \alpha v$ be put for y. Hence

$$\frac{d\phi}{du} - \frac{d}{dx} \frac{d\phi}{du'} + \frac{d^2}{dx^2} \frac{d\phi}{du''} - \ldots = 0.$$

The functions u and v are at present in our power; put $y - u$ for v and we have

$$\int \phi(x, y, y', y'', \ldots)\, dx$$

$$= \int \phi(x, u, u', u'', \ldots)\, dx$$

$$+ (y-u)\int_0^1 \left(\frac{d\phi}{du'} - \frac{d}{dx}\frac{d\phi}{du''} + \ldots\right) d\alpha$$

$$+ (y'-u')\int_0^1 \left(\frac{d\phi}{du''} - \ldots\right) d\alpha$$

$$+ \ldots\ldots\ldots\ldots\ldots\ldots\ldots\ldots$$

Thus $\int V dx$ is here actually exhibited as an expression consisting of terms, one involving only ordinary integration with respect to x, and the others ordinary integration with respect to α. The function u is still in our power; it should be chosen so that none of the quantities which occur become infinite or indeterminate; it may happen that consistently with this limitation we may put $u = 0$.

367. It will now be easy to give the necessary and sufficient conditions for ensuring that a function shall be integrable *per se* more than once.

Let V have the same meaning as before.

We have, whatever V may be,

$$\int \left\{ \int V\, dx \right\} dx = x \int V\, dx - \int x V\, dx.$$

In order then that V may be integrable *per se* twice, the condition must of course be satisfied which ensures that it is integrable *per se* once; and then the only additional condition is that xV must also be integrable *per se* once. Thus in order that V may be integrable *per se* twice, the necessary and sufficient conditions are that the following relations must be identically true,

$$\frac{dV}{dy} - \frac{d}{dx}\frac{dV}{dy'} + \frac{d^2}{dx^2}\frac{dV}{dy''} - \ldots = 0 \ldots\ldots\ldots\ldots (1),$$

$$\frac{dVx}{dy} - \frac{d}{dx}\frac{dVx}{dy'} + \frac{d^2}{dx^2}\frac{dVx}{dy''} - \ldots = 0 \quad \ldots\ldots\ldots (2).$$

We may modify the form of (2). For

$$\frac{dVx}{dy} = x\frac{dV}{dy}, \quad \frac{dVx}{dy'} = x\frac{dV}{dy'}, \quad \frac{dVx}{dy''} = x\frac{dV}{dy''}, \ldots\ldots;$$

$$\frac{d}{dx}\frac{dVx}{dy'} = x\frac{d}{dx}\frac{dV}{dy'} + \frac{dV}{dy'},$$

$$\frac{d^2}{dx^2}\frac{dVx}{dy''} = x\frac{d^2}{dx^2}\frac{dV}{dy''} + 2\frac{d}{dx}\frac{dV}{dy''},$$

$$\frac{d^3}{dx^3}\frac{dVx}{dy'''} = x\frac{d^3}{dx^3}\frac{dV}{dy'''} + 3\frac{d^2}{dx^2}\frac{dV}{dy'''},$$

$$\ldots\ldots\ldots\ldots\ldots\ldots\ldots\ldots$$

Substitute in (2) and omit the terms which are zero by (1); then we obtain

$$\frac{dV}{dy'} - 2\frac{d}{dx}\frac{dV}{dy''} + 3\frac{d^2}{dx^2}\frac{dV}{dy'''} - \ldots = 0 \quad \ldots\ldots\ldots (3).$$

Thus (1) and (2) may be replaced by (1) and (3).

By a formula given in Art. 54 the nth integral of any proposed expression is exhibited in terms of $n+1$ single integrals. From this formula we infer that in order that V may be integrable *per se* n times, it is necessary and sufficient that each of the following expressions should be integrable *per se* once,

$$V, \quad xV, \quad x^2V, \ldots\ldots x^nV.$$

For example, in order that V may be integrable *per se* three times, besides the conditions (1) and (2) or (1) and (3), the following must be identically true,

$$\frac{dVx^2}{dy} - \frac{d}{dx}\frac{dVx^2}{dy'} + \frac{d^2}{dx^2}\frac{dVx^2}{dy''} - \ldots = 0 \quad \ldots\ldots\ldots (4).$$

We may modify the form of (4). For

$$\frac{d}{dx}\frac{dVx^2}{dy'} = x^2\frac{d}{dx}\frac{dV}{dy'} + 2x\frac{dV}{dy'},$$

$$\frac{d^2}{dx^2}\frac{dVx^2}{dy''} = x^2\frac{d^2}{dx^2}\frac{dV}{dy''} + 4x\frac{d}{dx}\frac{dV}{dy''} + 2\frac{dV}{dy''},$$

$$\frac{d^3}{dx^3}\frac{dVx^2}{dy'''} = x^2\frac{d^3}{dx^3}\frac{dV}{dy'''} + 6x\frac{d^2}{dx^2}\frac{dV}{dy'''} + 6\frac{d}{dx}\frac{dV}{dy'''},$$

....

Substitute in (4) and omit the terms which are zero by (1) and (3); then we obtain

$$\frac{dV}{dy''} - \frac{3.2}{1.2}\frac{d}{dx}\frac{dV}{dy'''} + \frac{4.3}{1.2}\frac{d^2}{dx^2}\frac{dV}{dy''''} - \ldots = 0 \;\ldots\ldots\; (5).$$

Thus (5) may be taken instead of (4), in conjunction with (1) and (2) or (1) and (3).

Addition on the Variability of Limits.

368. In the method we have adopted of treating problems involving changes of the limits we have followed the example given in two most elaborate works on the subject, those of Strauch and Jellett; and we decidedly recommend this method as the best. We do not ascribe any *variation* to the independent variable, but only to the dependent variable. Another method however has been frequently adopted, and it should be explained in order that the student may understand any reference to it which may occur in his reading.

In this method a variation is ascribed both to the dependent and independent variables.

Let x become $x + \delta x$ and let y become $y + \delta y$, it is required to find the variations of $\frac{dy}{dx}$, $\frac{d^2y}{dx^2}$, ...

We denote the variation in $\frac{dy}{dx}$ by $\delta\frac{dy}{dx}$; thus

$$\delta\frac{dy}{dx}=\frac{d(y+\delta y)}{d(x+\delta x)}-\frac{dy}{dx}$$

$$=\frac{\frac{dy}{dx}+\frac{d\delta y}{dx}}{1+\frac{d\delta x}{dx}}-\frac{dy}{dx}$$

$$=\frac{dy}{dx}+\frac{d\delta y}{dx}-\frac{dy}{dx}\frac{d\delta x}{dx}-\frac{dy}{dx},$$

neglecting small quantities of the second order.

Thus adopting the usual notation for a differential coefficient, we have

$$\delta y'=\frac{d\delta y}{dx}-y'\frac{d\delta x}{dx}$$

$$=\frac{d(\delta y-y'\delta x)}{dx}+y''\delta x,$$

or

$$\delta y'-y''\delta x=\frac{d(\delta y-y'\delta x)}{dx}.$$

In this result change y into y'; thus

$$\delta y''-y'''\delta x=\frac{d(\delta y'-y''\delta x)}{dx}$$

$$=\frac{d^2(\delta y-y'\delta x)}{dx^2}.$$

Similarly $\delta y'''-y''''\delta x=\dfrac{d^3(\delta y-y'\delta x)}{dx^3}$,

and so on.

Put ω for $\delta y-y'\delta x$; thus

$$\delta y'=\frac{d\omega}{dx}+y''\delta x,$$

$$\delta y''=\frac{d^2\omega}{dx^2}+y'''\delta x,$$

$$\delta y'''=\frac{d^3\omega}{dx^3}+y''''\delta x,$$

....

Now let V be any function of x, y, and the differential coefficients of y with respect to x; and let $U=\int_{x_0}^{x_1} V\,dx$. Let it be required to express the variation of U which arises from the variations δx and δy in x and y respectively. Let δV denote the change made in V; then

$$\delta U=\int_{x_0}^{x_1}(V+\delta V)\frac{d(x+\delta x)}{dx}\,dx-\int_{x_0}^{x_1}V\,dx$$

$$=\int_{x_0}^{x_1}V\frac{d\delta x}{dx}\,dx+\int_{x_0}^{x_1}\delta V\,dx,$$

neglecting a term of the second order.

Now $$\int V\frac{d\delta x}{dx}\,dx=V\delta x-\int\left[\frac{dV}{dx}\right]\delta x\,dx,$$

therefore $$\int_{x_0}^{x_1}V\frac{d\delta x}{dx}\,dx=(V\delta x)_1-(V\delta x)_0-\int_{x_0}^{x_1}\left[\frac{dV}{dx}\right]\delta x\,dx,$$

where $\left[\frac{dV}{dx}\right]$ denotes the complete differential coefficient of V with respect to x.

Thus $$\delta U=(V\delta x)_1-(V\delta x)_0+\int_{x_0}^{x_1}\left\{\delta V-\left[\frac{dV}{dx}\right]\delta x\right\}dx.$$

And $$\delta V=\frac{dV}{dx}\delta x+\frac{dV}{dy}\delta y+\frac{dV}{dy'}\delta y'+\frac{dV}{dy''}\delta y''+\dots,$$

$$\left[\frac{dV}{dx}\right]=\frac{dV}{dx}+\frac{dV}{dy}y'+\frac{dV}{dy'}y''+\frac{dV}{dy''}y'''+\dots;$$

thus

$$\delta V-\left[\frac{dV}{dx}\right]\delta x=\frac{dV}{dy}\omega+\frac{dV}{dy'}\omega'+\frac{dV}{dy''}\omega''+\dots,$$

and finally

$$\delta U=(V\delta x)_1-(V\delta x)_0+\int_{x_0}^{x_1}\left(\frac{dV}{dy}\omega+\frac{dV}{dy'}\omega'+\frac{dV}{dy''}\omega''+\dots\right)dx.$$

We need not proceed further as we have arrived at a result equivalent to that in Art. 344; we have here ω instead of the

δy which occurs there, and δx_1 and δx_0 for dx_1 and dx_0 respectively.

In geometrical applications it will be observed that x and y become by variation $x + \delta x$ and $y + \delta y$ respectively. Thus $x_1 + \delta x_1$ will correspond to the $x_1 + dx_1$ of Art. 345, and $y_1 + \delta y_1$ will correspond to the $\left(Y + \frac{dY}{dx} dx_1\right)_1$ of Art. 345.

369. For further information on the Calculus of Variations the student may consult Professor Jellett's treatise, and the *History of the Progress of the Calculus of Variations during the Nineteenth Century*, by the present writer.

The most interesting examples in this subject are those which are connected with physical science, as the problem of the brachistochrone; accordingly we shall include some more applications of this kind in the following selection for exercise.

EXAMPLES.

1. A curve of given length has its extremities on two given intersecting straight lines; determine its form when the area included between the curve and its chord is a maximum.

2. Determine a plane closed curve of given perimeter which shall include a maximum area.
(See *History...*, page 68.)

3. Required to connect two fixed points by a curve of given length so that the area bounded by the curve, the ordinates of the fixed points, and the axis of abscissæ shall be a maximum, supposing the given length greater than is consistent with the solution in Art. 353.
(See *History...*, page 427.)

4. A rectangular dish is to be fitted with a tin cover of given height having the ends vertical; determine the form so that the amount of material used may be the least possible.

5. A mountain is in the shape of a portion of a sphere, and the velocity of a man walking upon it varies as the height above the horizontal great circle of the complete sphere; shew that if he wishes to pass from one point to another in the shortest possible time, he must walk in the vertical plane which contains the two points.

6. When a curved surface can be divided by a plane into two symmetrical portions the intersection of the plane and surface, when an intersection exists, is in general a line of minimum length on the surface.

(See *History*..., page 365.)

7. Find the minimum value of

$$\int \left\{ \left(\frac{dy}{dx}\right)^2 \sin x + \frac{(y + x - \sin x)^2}{\sin x} \right\} dx.$$

(See *Philosophical Magazine* for December, 1861.)

8. Required the minimum value of $\int_0^1 \left(\frac{dy}{dx}\right)^2 dx$ under the following conditions; $y_0 = 1$, $\int_0^1 \frac{y}{y_1} dx = -1$.

(See *History*..., page 432.)

9. Required the variation of $\int V dx$, where V is a function of $x, y, \frac{dy}{dx}, \frac{d^2y}{dx^2}, \ldots$ and v, where $v = \int V' dx$, and V' is also a function of $x, y, \frac{dy}{dx}, \frac{d^2y}{dx^2}, \ldots$

(See *History*..., page 21.)

10. Let s denote $\int_0^x \sqrt{(1 + p^2)}\, dx$, and let $\phi(s)$ be any function of s; then the relation between x and y is required which makes $\int_0^a \phi(s)\, dx$ a maximum or a minimum while $\int_0^a \sqrt{(1 + p^2)}\, dx$ has a given value, a being a constant. For a particular case suppose $\phi(s) = s$.

(See *History*..., page 453.)

11. Required the curve at every point of which

$$\left\{y + (m - x)\frac{dy}{dx}\right\}\left\{y + (n - x)\frac{dy}{dx}\right\}$$

is a maximum or a minimum.

(See *History*..., page 1.)

12. Required the curve at every point of which $y\frac{dy}{dx}$ is a maximum or a minimum, the variations of y and $\frac{dy}{dx}$ being so taken that at any point $yx - y^2\frac{dx}{dy}$ shall undergo no change by variation.

(See *History*..., page 414.)

13. Apply Art. 350 to prove the point assumed in Art. 339, namely, that the required curve in the brachistochrone problem lies in the vertical plane which contains the two given points.

14. The form of a homogeneous solid of revolution of given superficial area, and described upon an axis of given length, is such that its moment of inertia about the axis is a maximum; prove that the normal at any point of the generating curve is three times as long as the radius of curvature.

15. A given volume of a given substance is to be formed into a solid of revolution, such that the time of a small oscillation about a horizontal axis perpendicular to the axis of figure may be a minimum; determine the form of the solid.

(See *History*..., page 391.)

16. A vessel of given capacity in the form of a surface of revolution with two circular ends, is just filled with inelastic fluid which revolves about the axis of the vessel, and is supposed to be free from the action of gravity. Investigate the form of the vessel that the whole pressure which the fluid exerts upon it may be the least possible, the magnitudes of the circular ends being given.

Result. The generating curve is a catenary.

17. Find the equation given by the Calculus of Variations for the transverse section of a straight and uniform canal, when one of the three quantities, the surface, the capacity, and the normal hydrostatic pressure, is either a maximum or a minimum, and the other two are given, the terminal surfaces and pressures not being taken into account.

Shew also that when the surface is a minimum and the capacity only is given, the section is circular; and when the normal pressure is a minimum the section is a catenary or two straight lines, according as the surface or the capacity is given.

18. If there are two curves with their concavities downwards and terminated in the same extremities, a particle moving under the action of gravity will take a longer time to describe the upper curve than the lower curve, the initial velocity being supposed the same in the two cases.

(See *History*..., page 348.)

CAMBRIDGE: PRINTED AT THE UNIVERSITY PRESS.

MATHEMATICAL WORKS

BY

I. TODHUNTER, M.A.

Fellow and Principal Mathematical Lecturer of St John's College, Cambridge.

1. A Treatise on Algebra, for the use of Colleges and Schools; with numerous Examples. Second Edition, revised. Crown 8vo, cloth, 7*s*. 6*d*.

2. An Elementary Treatise on the Theory of Equations, with a Collection of Examples. Crown 8vo, cloth, 7*s*. 6*d*.

3. A Treatise on Plane Trigonometry, for the use of Colleges and Schools; with numerous Examples. Second Edition, revised. Crown 8vo, cloth, 5*s*.

4. A Treatise on Spherical Trigonometry, for the use of Colleges and Schools; with numerous Examples. Crown 8vo, cloth, 4*s*. 6*d*.

5. A Treatise on Plane Co-ordinate Geometry, as applied to the Straight Line and the Conic Sections; with numerous Examples. Third Edition. Crown 8vo, cloth, 7*s*. 6*d*.

6. A Treatise on the Differential Calculus; with numerous Examples. Third Edition, revised. Crown 8vo, cloth, 10*s*. 6*d*.

Mathematical Works by I. Todhunter, M.A.

7. A Treatise on the Integral Calculus and its Applications; with numerous Examples. Second Edition, revised and enlarged. Crown 8vo, cloth, 10*s.* 6*d.*

8. A Treatise on Analytical Statics; with numerous Examples. Second Edition, revised and enlarged. Crown 8vo, cloth, 10*s.* 6*d.*

9. Examples of Analytical Geometry of Three Dimensions. Crown 8vo, cloth, 4*s.*

10. A History of the Progress of the Calculus of Variations during the Nineteenth Century. 8vo, cloth, 12*s.*

MACMILLAN AND CO. CAMBRIDGE AND LONDON.

Printed in Dunstable, United Kingdom

91219939R00201